THE MARCH OF UNREASON

TO JANICE

THE MARCH OF UNREASON

Science, Democracy, and the New Fundamentalism

DICK TAVERNE

OXFORD
UNIVERSITY PRESS

OXFORD
UNIVERSITY PRESS

Great Clarendon Street, Oxford OX2 6DP

Oxford University Press is a department of the University of Oxford.
It furthers the University's objective of excellence in research, scholarship,
and education by publishing worldwide in

Oxford New York

Auckland Cape Town Dar es Salaam Hong Kong Karachi
Kuala Lumpur Madrid Melbourne Mexico City Nairobi
New Delhi Shanghai Taipei Toronto

With offices in

Argentina Austria Brazil Chile Czech Republic France Greece
Guatemala Hungary Italy Japan South Korea Poland Portugal
Singapore Switzerland Thailand Turkey Ukraine Vietnam

Oxford is a registered trade mark of Oxford University Press
in the UK and in certain other countries

Published in the United States
by Oxford University Press Inc., New York

British Library Cataloguing in Publication Data

Data available

Library of Congress Cataloging in Publication Data

Data available

ISBN 0-19-280485-5

1

Typeset by RefineCatch Limited, Bungay, Suffolk
Printed in Great Britain by
Clays Ltd, St Ives plc

Contents

Acknowledgements

As a layman writing about specialized topics, I have been hugely dependent on advice from experts. Others have helped with more general comments on the book or in other important ways. I cannot thank them warmly enough for their invaluable support and encouragement. They are not of course responsible for my errors and misjudgements.

I cannot name them all, but they include: Wilfred Beckerman, Tracey Brown, Nick Bunnin, Adam Burgess, Peter Campbell, Gordon Conway, Buck Creel, Andrew Cockburn, Bill Durodié, John Emsley, Marsha Filion, Mike Fitzpatrick, Walter Gratzer, Abbie Headon, Stephen Hearst, David Henderson, Sally Hirst, Roger Kalla, Chris Leaver, Bryan Magee, Mark Matfield, Latha Menon, Bill Newton-Smith, Bridget Ogilvie, Hugh Purcell, Michael Rodgers, Hilary Rubinstein, Neil Summerton, Ray Tallis, Tony Trewavas, the late Bernard Williams and Lewis Wolpert.

Prologue

THIS book is about science and society. Since I am neither a scientist nor a sociologist, but a former lawyer and politician with some experience of government and industry, perhaps I should explain why I have wandered into unfamiliar territory.

I am married to a biologist and I have long been acutely aware how little most people know about science. What I find especially disturbing is that some people not only do not know about science, but do not want to know and seem proud of not knowing. Yet science, especially the science concerned with health and the environment, has come to play an ever greater part in our lives.

Like many others, I fell under the spell of Rachel Carson when I read *The silent spring* soon after it was published in 1962. I was persuaded that the threat which technology posed to the environment should be taken far more seriously than it was and started to read books by Paul Ehrlich and Barry Commoner, who were telling us about the disasters that lay ahead. In the late 1960s, when I was a Treasury Minister, I took time off from contemplating the economic problems of the UK to attend a conference at which Paul Ehrlich was the star attraction. I was duly impressed by his eloquent prophesies of doom, delivered with a kind of cheerful resignation ('If you are travelling on the *Titanic*, you may as well travel first-class'), but I also noted the somewhat less cataclysmic views of another scientist, a wise man called Kenneth Mellanby, who argued that while there were grounds for concern, it was unlikely that we would in fact starve or be poisoned or run out of energy or other vital resources as Ehrlich predicted. A few years later the Club of Rome published *The limits to growth*, which claimed that economic growth would have to stop as the world was running out of resources. I was still sufficiently in thrall to the fashionable doomsters to believe that, unless we radically changed our ways, our quality of life could not survive. I joined Friends of the Earth and Greenpeace. Indeed, I would pay tribute to the useful service both performed in their early days in rousing public opinion from a certain smug indifference to the dangers of environmental degradation.

In the mid-1970s, to make our small contribution to cleaner air, my wife and I decided to give up owning a car (which was easy for us, as we live in central London) in favour of bicycles. Incidentally, whatever its environmental merits, the decision proved extremely convenient. A bicycle has been my main form of urban transport for over thirty years and I have become more convinced than ever about its virtues. It is a most enjoyable way to travel about London. You can be sure of arriving on time; you suffer none of the frustrations of being stuck in traffic jams and not finding anywhere to park; you do not have to worry about dents or scratches on your car; and it is much healthier than motoring. People worry about safety, but a comparison on an actuarial basis of 'life years' lost through cycle accidents with gains from improved fitness reveals that for every life year lost through accidents, twenty are gained from improved health.[1] The bicycle is also one of the most efficient machines ever invented for converting energy into motion: it has been described as a 'green' car, which 'runs on tap water and toasted teacakes, and has a built-in gym'.[2] But most important of all, the quality of urban life would be greatly improved if many more journeys were made by bicycle. There is no reason why this aim cannot be achieved in the UK. In Denmark, for instance, as a result of careful planning, more than 20 per cent of all journeys are made by bicycle; in Britain the figure is 3 per cent. Yet the Danes own more cars per head than the British.

I cite my devotion to the bicycle as evidence that when I criticize the excesses of some environmentalists it is not because I do not regard care for the environment as one of the important issues of our time. But I am a pragmatic environmentalist. Risk must be weighed against benefit. I want analysis of the risk of damage to the environment to be based on evidence and recommendations for remedial action to be based on science rather than emotion. I care not only about the environment but about reason.

Human beings have developed this wonderful gift and constantly ignore it. Just as we learn more about our genetic make-up and find better ways of dealing with deadly diseases, more people turn to homeopathy and other quack remedies. When it comes to food and farming, the voice of reason is stilled and the public turns to a vague yearning to go 'back-to-nature'. Religious fundamentalism is rampant, not only in Islam and among Jewish settlers in Palestine; in America we witness the spread of creationism and the return to the beliefs that prevailed before the Enlightenment banished superstition and modern science was born. Millions of born-again Christians believe in a primitive religion that features an interventionist God who, it seems, periodically answers prayers to help but is never the cause of harm. To cite one example that is not atypical: when interviewed after the hijack of an American plane, the pilot thanked God for answering his prayers and bringing him safely through his ordeal. It did not occur to him that God had also answered the prayers of the devout Muslim hijackers and helped them to seize the plane. I reflected, somewhat irreverently, that his God had much in common with the late Lord Mountbatten, who eventually became Viceroy of India and Chief of Defence Staff. In his earlier life he was an intrepid young naval commander in the Second World War, of whom his naval colleagues said: 'No one like Dickie Mountbatten to have with you in a tight spot. No one like Dickie to get you into one.' The pilot's gratitude for divine intervention would be a matter of private belief and of no particular importance, were it not for the growing influence of religious fundamentalism. Such fundamentalism is a serious danger to peace and democracy. It spreads intolerance wherever it is found.

Optimism about scientific progress faded some time during the last century. Today science and reason are under siege from many quarters. Many people have become increasingly sceptical about the benefits of new technology and no longer trust experts. Possible risks from new developments loom larger in the public mind than possible benefits and we hear constantly about the need to apply 'the Precautionary Principle', as if it is some scientific law that needs no further explanation. (Indeed, when it is carefully ana- lysed, it turns out to be either trite, or meaningless, or positively harmful.) At the same time, it is fashionable in some academic circles to question the objectivity of science, to argue that what matters is the values of scientists rather than their findings, and indeed to doubt whether any truths can be regarded as objectively established. I do not share this pessimistic, indeed one might call it nihilistic, view. I agree with the American philosopher C. S. Peirce: 'A man must be downright crazy to doubt that science has made many true discoveries'. Individual scientists may err or be influ- enced by their prejudices, but the scientific process is essentially a communal and iterative process, in which each constantly checks his or her own and others' mistakes until some sort of objective view emerges. The great virtue of science is that its truths must be reproducible and are independent of time, place, and personality.

Gradually, as I began to look more critically at the attitudes to science of the Green activists and the more passionate environ- mentalists, I found that passion (including a passion for publicity) tends to prevail over reason and regard for evidence. *Limits to growth* was shown to be based on erroneous assumptions. A new eco-fundamentalism has emerged, with a powerful influence on policy. In 2001 when a Danish statistician, Bjorn Lomborg, pro- duced facts and figures that presented a strong prima facie case against the belief of many environmentalists that the world is facing an impending debacle,[3] he was answered, not by carefully marshalled evidence and arguments, but by a torrent of abuse. One of his opponents threw a pie in his face and others applauded. He was regarded as a heretic who had dared to question their religion.

One person who persuaded me more than anyone else to ques- tion claims of approaching doom was, ironically, the American activist Jeremy Rifkind. In the 1980s, he was the most vociferous opponent of genetic engineering (the term then generally used where genetic modification is used today). He accused scientists of playing 'ecological roulette' and predicted catastrophic consequences from the release of thousands of genetically engineered organisms into the environment. Even if the chances that any one of the new organisms would run amok were remote, he argued that by 'sheer statistical probability' some of them were bound to prove disastrous. The most dramatic of his many claims of impending doom was that the introduction of an 'ice-minus' bacterium into such plants as potatoes or strawberries to protect them against frost would alter rainfall patterns and cause global drought. The claims were thoroughly tested by the courts, the US Environment Protection Agency, and the former Office of Technology Assessment of Congress and were found to be without any foundation.[4] He was the intellectual version of a sandwich-board man patrolling Oxford Street with the warning: 'The end of the world is nigh'. None of his dire predictions have materialized, but he continues to be treated as an eminent authority by the media and is still regarded as a guru by eco-warriors.

In the late 1980s and early 1990s, fears about genetic modification were much more widespread on the continent of Europe than in Britain. In Germany its extreme opponents fire-bombed one of the Max Planck Institutes because it was conducting genetic research on petunias. They argued that as genetic modification was bound to lead to eugenics, and as this had been practised by the Nazis, such research was bound to lead to Nazism. An expensive plant built by Hoechst to manufacture recombinant human insulin in bacteria stood idle for years because of threats from anti-GM campaigners; this hormone has since proved of enormous benefit to sufferers from diabetes. In Britain on the other hand, polls published at that time showed that fears about the new biotechnology were restricted to a small and rather ineffective minority.

However, attitudes in Britain have changed. This is partly the result of a number of public disasters of which much the most influential was the traumatic experience of BSE (bovine spongiform encephalopathy) that undermined trust in experts. It was partly the success of Greenpeace and Friends of the Earth in exploiting a series of scare stories. In the early 1990s, for example, Greenpeace fought a campaign against a proposal to install a brand-new, state-of-the-art incineration plant in Cleveland, in the north of England. The plant was designed to provide better ways of disposing of toxic industrial waste than by dumping it in landfill. Greenpeace distributed leaflets alleging that an incineration plant would cause cancer by releasing dioxins, however small the amount. The slogan was STINC: 'Stop Incineration in Cleveland'. The local population was roused to vigorous demonstrations and the campaign was totally successful. The plant was never built and toxic chemicals continued to be deposited in landfill sites instead of being rendered harmless through incineration. In fact the amount of dioxins released into the atmosphere would have been minute, well below any conceivable danger level. Green lobbyists continue to oppose every proposal to build incineration plants and, when asked what should be the alternative, answer: 'All waste should be recycled' or 'We must stop creating any waste'. It is the age-old cry of the millennialist: nothing is worth doing until we have built Jerusalem.

Greenpeace had its greatest success with its *Brent Spar* campaign in 1995. *Brent Spar* was a disused giant oil-rig owned by Shell, who had decided, after careful consultation about the environmental effects, to dispose of it in the deep waters of the mid-Atlantic. Greenpeace organized an extremely effective Europe-wide boy-cott of Shell petrol stations to protest against the company's plans to pollute the ocean. For days on end, Greenpeace dominated TV news bulletins throughout Europe with shots of brave warriors in their small inflatables harassing and trying to stop huge tugs towing the rig.

The campaign was a triumph. One of the world's most powerful companies was forced into a humiliating climb-down and had to order the tugs to turn round and leave *Brent Spar* in a Norwegian fjord instead. From an environmental point of view, the campaign was misconceived and, like the campaign against incinerators, ignored scientific evidence. Claims made by Greenpeace that the rig was full of toxic residues were shown to be entirely without foundation—indeed Greenpeace wrote to Shell apologizing for the factual error. Furthermore, disposal in mid-Atlantic would have provided an attractive underwater playground for a variety of fish and would have been a much cheaper and environmentally more beneficial way of disposing of the rig, as was later confirmed by the Natural Environmental Research Council. Indeed I believe that there must be considerable doubts about the Greenpeace belief in its own propaganda. What the public did not know and Greenpeace did not mention was that, when its own ship *Rainbow Warrior* was irreparably damaged by French saboteurs in New Zealand in 1986, Greenpeace deliberately sank it off the coast of New Zealand and claimed that it would form an artificial reef that would be of great benefit to marine life.[5] Since then, in 1998, Greenpeace campaigned, again successfully, for a ban on all marine disposal of disused oil installations.

The key battleground on which the forces of science and anti-science now clash is the future of genetically modified crops. The issue itself is not only of great importance to the future of agriculture and the environment, especially in the developing world, but it is of central importance to the theme of this book, because it symbolizes the conflict between the evidence-based approach and dogma. Genetic modification is to Greenpeace, Friends of the Earth, and kindred organizations, what abortion is to Roman Catholics and American evangelicals. Evidence, if any, is cited not in the pursuit of truth but to support passionately held beliefs. In the debates in Britain about stem cell research, Catholics discussed the scientific issues, but without exception argued that adult stem cells could be used for research just as effectively as those from embryos, despite the balance of evidence on the other side. To Catholics, the use of embryo stem cells could not be allowed to be more effective because their use was contrary to religious dogma. They would not allow it to be possible that evidence might change their minds. To many of the Green lobbies rejection of GM technology has become a tenet of faith, and any evidence that contradicts the faith is simply irrelevant.

What makes the attitudes of the Green lobbies a matter of

special concern is the contrast between the treatment by the media of ordinary party politics and of green issues. In my political career I have found that politicians get a worse press than they generally deserve, except in one respect: many a good man and woman has been corrupted by the demands of party loyalty. In the culture of the British Parliamentary system, which tends to be shared by lobby correspondents reporting on Parliament, those who sacrifice their personal principles to stay loyal to their party are on the whole regarded as virtuous. They have done the right thing. On the other hand, those who abandon their party to stay loyal to their principles are regarded, certainly by former colleagues, as traitors. 'Damn your principles,' said Disraeli. 'Stick to your party.' Tribalism rules and principles may not be allowed to challenge its sovereignty. Too many politicians forget that parties are created to pursue particular aims and express particular principles, and that parties in themselves, if they abandon these principles, have no particular virtue and deserve no irrevocable loyalty.

Reason too becomes a casualty of tribalism. Party spokesmen will argue a case about which they have private misgivings because it suits party interests. Indeed extreme Opposition spokesmen will blame the government of the day for every conceivable mishap and hold it responsible for the caprice of nature as well as the follies of man. It was in revolt against this ethos that towards the end of my relatively brief career in the House of Commons I left my party and was twice re-elected as an independent MP. Of course democratic politics are meaningless without parties, but parties can survive without tribalism; indeed tribalism and excessive partisanship undermine democracy.

Green lobbies are, if anything, even more ready to sacrifice reason for the sake of dogma than politicians are for the sake of party. Weighty reports from authoritative sources that have no axe to grind, which show that GM crops can offer substantial potential benefits to the developing world and that there is no special reason to suppose they are dangerous to human health, are simply ignored. Flimsy evidence from highly partisan sources (seldom if ever peer-reviewed), which appears to support their case against

GM crops, is uncritically accepted. Just as parties are a necessary part of democracy, environmental lobbies play an important part in making people and governments aware of environmental issues. But blind loyalty to the cause is just as corrupting as tribalism in party politics. In fact it is more dangerous, because the media subject the pronouncements of parties to ruthless criticism, but treat environmental groups like The Soil Association, Greenpeace, and Friends of the Earth as independent authorities above criticism, as if they were a sort of collective Mother Theresa.[6] There is a general feeling that, since they are trying to save the planet, they must be right. This enables them to make statements that ignore evidence about the effects of genetic modification, or for that matter about the polluting effects of old warships or disused oil rigs or pesticide residues, that go largely unquestioned and uncontradicted.

So far the campaign against GM crops by Green lobbies has been very successful. It has won wide public support in Europe and has effectively undermined an important technology. The influence of 'green' non-governmental organizations, or NGOs, has increased, and is increasing, throughout the European Union. Governments treat them as official representatives of consumer opinion and they are to be found at the heart of policy formulation.

I regard their increasing influence as deeply disturbing. They exploit the media brilliantly and have managed to convey the impression that they are a noble band of crusaders struggling against malign forces in society that will damage or destroy the planet. They foster public suspicion about science and mistrust of experts and have succeeded in driving scientists onto the defensive. A mood has been created in which scientists themselves have come to feel that somehow public ignorance of science, indeed public suspicion of science, is their own fault.

In my view, the lack of public understanding of science and the apparent lack of concern of the public for the evidence-based approach should concern non-scientists more than it does. My theme is that reliance on dogma and ideology instead of evidence is unhealthy for democracy. Reason is one of the foundations of democracy. If irrationality prevails and respect for evidence is rejected, how can we resist religious fundamentalism and chauvinism and racism and all the other threats to a civilized society? We become a credulous society ready to believe charlatans and risk sinking back into superstition and the savagery that prevailed before the Enlightenment. The building blocks of today's liberal democracies were laid in the seventeenth and eighteenth centuries, in the period celebrated by Roy Porter in his wonderful book *Enlightenment Britain and the creation of the modern world*. It is no coincidence that this was the time when modern science was born. Indeed science was the chief progenitor of the Enlightenment. Both science and democracy are based on the rejection of dogmatism, and whenever and wherever ideology rules, freedom as well the evidence-based approach is suppressed.

I do not suggest that there is a lack of public interest in science. There is a plethora of books and articles that clearly explain the latest developments in non-technical terms. Books about science have never been more popular; but few writers are concerned about the wider implications for society of rejecting the scientific method. I also believe that there is room for a non-scientist to sing the praises of science as one of the glories of mankind and to defend scientists against the mistaken, often bizarre, charges made against them.

Of course, there are grave risks for any lay person who trespasses on professional territory. This applies not only to discussion of the latest developments in plant breeding, toxicology, medicine, and other aspects of environmental science, but also of the attacks made on scientific truth by postmodernist philosophers and sociologists. But I believe non-scientists (and non-philosophers and non-sociologists) like myself should be able to distinguish obviously bogus from valid arguments and to judge between claims based on careful assessment of evidence and manifestations of a sham reasoning, which uses evidence selectively and unscrupulously to bolster prejudice and goes through the motions of inquiry only to demonstrate some foregone conclusion. I also regret the compartmentalization of intellectual disciplines, which leaves discussion of some subjects either to experts, many of them talking to each other, or to professional commentators, the village pundits of the press, offering their pearls of wisdom for the edification of the populace.

I believe non-scientists and especially politicians who are concerned about the interaction of science with society should take special care to try to understand and evaluate scientific evidence about controversial questions of the day. Is there any reason to have recourse to alternative medicine? Is 'organic' farming really a better alternative to conventional farming or the cultivation of transgenic crops, and do government subsidies for organic farmers have any possible justification? Are transgenic crops in fact a threat to our health or to the environment? Can they reduce hunger, disease and environmental degradation? Are there rational grounds for the popular fear that science may have over-reached itself, or for the claims of pessimists that only a dramatic and revolutionary transformation of western society and culture can save the world for future generations? Are technological developments exposing us to unacceptable risks, so that we should apply the Precautionary Principle to new developments? These questions are not Eleusinian mysteries that can only be understood by initiates. They are questions about which people in public life should be able to express an informed view.

They have become intensely political questions, especially as they often involve multinational companies. Suspicion of science is mixed up with a new anti-capitalist mood and the anti-science movement today regards itself as left wing, whereas traditionally it was the left which linked science with progress and the right which preached a doctrine of 'back-to-nature' based on a rejection of science. In fact, in arguments for and against particular scientific developments or about science and society, distinctions between left and right are meaningless. What is at stake is the role of reason in democracy. What is also at stake is truth. Most newspapers in Britain do not give accuracy in reporting as high a priority as newsworthiness, with the result that Green lobbies can make unsubstantiated statements in flagrant disregard of facts and be assured of huge coverage. Public misconceptions may be corrected in the end, but they can persist long enough to do immense damage.

Finally, what may also be at stake is the economic prosperity and quality of life in Britain and Europe. There is a danger, not imminent but not inconceivable, that our science and technology could decline into relative insignificance. There is no law which decrees that science must always flourish in Europe because Europe was the birthplace of modern science.

Between the eighth and thirteenth centuries in the golden age of Islam, Arab thinkers led the world in mathematics, chemistry, astronomy, and medicine. They also preserved for us the civilization of ancient Greece. Then, sometime in the fourteenth century, religious dogmatism suppressed their spirit of scientific inquiry. Printing presses, for example, were banned in case they undermined the Word of God as revealed in the Koran and other sacred texts and science never recovered its place of glory in the Islamic world. China provides another example of self-inflicted technological decline. By the early fifteenth century, Chinese technology was probably the most sophisticated in the world. Not only had the Chinese invented gunpowder, the compass, and printing, but they surpassed all others in the technology that could give them control of the seas: shipbuilding. Hundreds of ships up to 400 feet long, which dwarfed the puny ships of European nations, dominated the Indian Ocean. Then a faction came to power which dismantled shipyards and banned ocean-going ships so that no more ships were built that could challenge the rising power of the fleets of Europe.[7]

European civilization could suffer the same fate. Eco-fundamentalists who elevate dogma over evidence exercise great influence among Europe's green pressure groups. They may damage science as effectively in Britain and Europe as did the Islamic fundamentalists in the Arab world and anti-technology mandarins in medieval China. Already companies that advance agricultural biotechnology have largely abandoned their operations in Europe. It is likely that future research and development in agricultural science will be concentrated in the United States, China, and India, and perhaps in Brazil and Mexico. If animal rightists prevail, the pharmaceutical industry could join this exodus. Not only our economy, but the intellectual quality of European civilization will suffer if our science base is gradually eroded.

My book starts with the birth of modern science at the time of the Enlightenment in Britain, which was also the time when liberal democracy was born. The two were linked at birth. John Locke, who can justly be called the father of liberal democracy, explicitly acknowledged the influence of the new scientific approach to his political ideas. It was also a time of optimism about the role of science in improving the condition of mankind. I trace some of the reasons for the change from optimism to the widespread suspicion and pessimism towards science that exist today and identify the rise of the environmental movement as probably the most significant. There are three issues that illustrate current and prevalent discomfort about the impact of science on our relationship with nature, often expressed in sentiments that we interfere with nature at our peril. One is the fashion for homeopathy and alternative medicine. Another is the popularity of organic farming, which has no scientific basis for the claims made on its behalf. The third is the most important to my central theme: the environmentalists' rejection of genetically modified crops, the issue that inspires the most passionate argument between those who support the evidence-based approach and those whose opposition has become a matter of dogma. I review in some detail the arguments for and against GM crops.

Why has this dogmatism arisen? Why do some of the Green activists evoke fear and hysteria? One reason is that part of the environmental movement has become eco-fundamentalist and turned into a crusading movement with all the attributes of a new religious faith. Another manifestation of the mood of suspicion towards science is found in the frequent invocation of the so-called 'Precautionary Principle',[8] which both affects and exemplifies current attitudes to issues of scientific controversy and could prove to be a serious obstacle to innovation and the spirit of enterprise.

The main intellectual case against science and technology, which has also contributed to the march of unreason, is the assault by postmodernists and relativists on the very citadel of science itself, its claim to objectivity and to being value-free. The main political case, particularly against biotechnology, is that it is promoted by multinational companies and that these villains are responsible for the menacing spread of globalization. The profit motive, it is often argued, corrupts science and causes bias in the results of research, while globalization increases poverty and inequality. I believe both arguments are largely misconceived. Finally, I return to the theme of science and democracy, to argue that despite the apparent irrationality of the democratic process, the two are interdependent and face common enemies: autocracy and fundamentalism, whatever form they take. Our willingness to accept evidence and to apply the evidence-based approach to the problems of government are ultimately issues that go to the heart of the nature of our society.

That is why, as a liberal democrat in politics, a pragmatic environmentalist, a non-scientist but a passionate believer in the importance of reason and truth, I felt compelled to write this book.

1

From Optimism to Pessimism

Nature and Nature's laws lay hid in night:
God said 'Let Newton be' and All was Light

Alexander Pope

'OF arms and the man I sing', wrote Virgil at the start of his epic about Aeneas and the founding of Rome. My theme is science and society or, more precisely, the importance of the evidence-based approach to a healthy democracy. Virgil's *Aeneid* started with tragedy—the fall of Troy—and ended with hope, the founding of Rome. My theme starts with the Enlightenment and the new optimism aroused by the birth of modern science and the first stirrings of democracy. But in the last century, optimism about science turned sour and today many new discoveries and technological developments are viewed with apprehension rather than hope. The new Rome that science built is under siege by the barbarians.

The Enlightenment was an extraordinary period. Isaiah Berlin called it one of the best and most hopeful episodes in the history of mankind, because, he wrote, 'the intellectual power, honesty, lucidity, courage and disinterested love of the truth of the most gifted thinkers of the 18th century remain to this day without parallel'.[1] He might have added that it was not just the eighteenth-century thinkers who deserved this accolade, but also some of the earlier ones of the seventeenth century. As Roy Porter has pointed out,[2] there has been a tendency to identify the Enlightenment with the eighteenth-century French *philosophes*, when some of the seventeenth-century thinkers in Britain, who were looked upon by the *philosophes* as their inspiration, exercised an influence that was at least as important. Voltaire, for example, in his *Lettres*, described England, perhaps over-generously, as a nation of philosophers and the cradle of liberty, tolerance, and sense. Francis Bacon, to him, was the prophet of modern science, Isaac Newton had revealed the laws of the universe, and John Locke had demolished Descartes and rebuilt philosophy on the bedrock of experience.[3] Denis Diderot (editor of the seminal reference text of the Enlightenment, the *Encyclopédie*) likewise acknowledged that 'without the English, reason and philosophy would still be in the most despicable infancy in France'.[4]

The Enlightenment in Britain, according to Roy Porter, made the world we have inherited, 'that secular value system to which most of us subscribe today which upholds the unity of mankind and basic personal freedoms, and the worth of tolerance, knowledge, education and opportunity.'[5] There was no special Enlightenment project; but there was a gradual revolution of ideas, which overturned years of sterile metaphysics, dethroned theocracy, saw the passing of the Divine Right of Kings, repealed the witchcraft statutes, introduced smallpox vaccination, ceased to treat infanticide as the product of bewitchment but as a crime, ceased to regard madness as a supernatural occurrence, but as an illness, and generally led to the withering of superstition under the light of reason. It was a period when political pamphlets sold tens of thousands of copies.[6] Indulgence in leisure and pleasure increased, with a new concern for happiness to which organized religion in its day of dominance had been inimical, and conspicuous delight was taken in food, helped by low prices and the introduction of such exotica as pineapples.[7] There was a new sense of optimism about the prospect, indeed some thought the inevitability, of progress, which contrasted with the gloomy view of theologians that the climate was deteriorating, the soil growing exhausted, and pestilences multiplying.

The birth of modern science

Central to these changes in attitudes was the birth of modern science, which in turn inspired the first tentative steps towards democracy. In pre-Enlightenment days, Calvin could claim to refute Copernicus with the text 'The world also is stablished, that it cannot be moved' (Psalms 93:1) adding 'Who will venture to place the authority of Copernicus above that of the Holy Spirit?' Galileo could be terrorized by the Inquisition into recantation. In fact, Galileo was one of the most important progenitors of the Enlightenment, not only because of his scientific discoveries, but because he dared to challenge authority and revelation as the source of knowledge. He asserted (until he was forced to retract) that the authority of the almighty church should have no right to interfere with the truth-seeking activities of science. 'Why', he said,

this would be as if an absolute despot, being neither a physician nor an architect, but knowing himself free to command, should undertake to administer medicines and erect buildings according to his whim—at grave peril of his poor patients' lives, and speedy collapse of his edifices.[8]

The challenge to the authority of the Catholic Church and the installation of reason in its place was the essential prelude to the birth of modern Western civilization.

In England, Newton ruled. Newton's *Principia* and Halley's calculations of the orbits of the comets were widely circulated and the laws of nature had established such a hold on intelligent men's imagination that, by the end of the seventeenth century, magic and witchcraft had become incredible. In Shakespeare's time, comets were still regarded as portents. After Newton, they were seen as being as obedient as the planets to the laws of gravitation. Newton as well as Galileo ensured that the validity of statements about the world no longer depended on the authority of those who made them but on the evidence in their support.

Perhaps it was Francis Bacon, however, who had the most profound and lasting influence on the Enlightenment and subsequent generations, because he, above all, can be regarded as the father of modern science. Bacon was no paragon of virtue. At the height of his career he was convicted of accepting bribes (not an altogether unusual practice at the time) and dismissed from office, which led him to concentrate on ways of advancing the cause of science. In Pope's words, he was 'the wisest, brightest, meanest of mankind'. He was a polymath, a man distinguished in politics, literature, philosophy and science, but his main profession was the law. Indeed, he rose to be Lord Chancellor. As his biographer John Henry observes, he made no new discoveries, developed no technical innovations, uncovered no previously hidden laws of nature,[9] yet his contribution was immense. Henry points to three key factors that account for his importance: an insistence on experimental method; a notion that a new knowledge of nature should be turned to the practical benefit of mankind; and the championing of inductive over deductive logic. He was acknowledged by the Royal Society, when it was founded some forty years after his death, as the father of experimental philosophy and its inspiration.

The birth of liberal democracy

One rival to Bacon's claim to be the most influential figure in the pantheon of the Enlightenment in Britain is John Locke. If Bacon is the father of modern science, Locke could reasonably be called the father of liberal democracy, the system of government in which sovereign power resides in the people, but where respect for the wishes of the majority is balanced by respect for the rule of law, human rights, and regard for the rights of minorities. Liberalism was not of course a British or a Lockean invention. It was a product of Britain and The Netherlands (the country in which both Locke and Voltaire had to seek refuge from domestic intolerance). The origins of liberalism were Protestant, but tolerant of other religions. Anglo-Dutch liberalism valued commerce and industry, had immense respect for the rights of property and supported freedom of expression. The Netherlands, the most advanced, libertarian, and egalitarian country in Europe of its day,[10] was the home of Baruch Spinoza, the first philosopher to articulate clearly the right to freedom of speech and to argue that it was a necessary means for securing public order. Nevertheless, Locke's philosophy set out more clearly than anyone before him, and many who have sought to follow him, some of the fundamental principles on which a liberal democracy must be based.

To start with, he was the great empiricist, deeply influenced by Bacon's scientific method. Indeed, he regarded his task and his philosophy as subservient to the role of scientists, the master builders

whose mighty designs, in advancing the sciences, will leave lasting monuments to the admiration of posterity ... Everyone must not hope to be a Boyle or a Sydenham and in an age which produces such masters as ... the incomparable Mr Newton ... it is ambition enough to be employed as an under-labourer in clearing the ground a little and removing some of the rubbish that lies in the way to knowledge.[11]

It was this regard for science that led him not to follow Descartes' road of rationalism and deduction but to base his philosophy, in Voltaire's words, 'on the bedrock of experience'. All our knowledge, he argued, except for logic and mathematics, is based on experience. He was contemptuous of metaphysics. Like Bacon, he rejected the idea that reasoning must be based on the Aristotelian syllogism, which 'has been thought more proper for the attaining victory in dispute than for the discovery or confirmation of truth in fair inquiries'. Hence his famous remark: 'God has not been so sparing to men to make them two-legged creatures, and left it to Aristotle to make them rational'.[12] Evidence, not prejudice, must be the basis of opinion:

... there are very few lovers of the truth, for truth's sake, even among those who persuade themselves that they are so. How a man may know whether he be so in earnest, is worth inquiry: and I think there is one unerring mark of it, viz. the *not entertaining any proposition with greater assurance than the proofs of it may warrant*[13] (My italics).

If Bacon was not one of the great scientists, though his influence was immense, Locke likewise was not one of the great philosophers. He was always ready to sacrifice logic whenever a logical argument appeared to conflict with common sense. But his occasional lapses of logic do not detract from his historic contribution, nor does the fact that many of his doctrines have dated. The two main reasons why Locke can legitimately be regarded as the father of liberal democracy are his total opposition to dogmatism and extremism and his recognition that the right of all individuals to own and use property subject to well-defined constraints of the law is the bedrock of a liberal society.

For Locke, truth was difficult to ascertain and therefore a rational being should hold his opinions with a measure of doubt. It followed that we should be wary of seeking to impose our opinions on others.

'Since ... it is unavoidable to the greatest part of men, if not all, to have several opinions, without certain and indubitable proofs of their truth; ... it would, methinks, become all men to maintain peace, and the common offices of humanity, and friendship, in the diversity of opinions; since we cannot reasonably expect that anyone should readily and obsequiously quit his own opinion, and embrace ours, with a blind resignation to an authority which the understanding of man acknowledges not. ... For where is the man that has incontestable evidence of the truth of all that he holds, or of the falsehood of all he condemns; or can say that he has examined to the bottom all his own or other men's opinions?'[14]

Again: 'Some eyes want spectacles to see things clearly and distinctly; but let not those that use them therefore say nobody can see clearly without them'.[15]

Locke's moral philosophy has also had a beneficial influence. His argument that the pursuit of happiness is the foundation of all liberty and that prudence is the most important of all virtues may seem a somewhat limited view of ethics. The emphasis on prudence was very much a reflection of the times, which saw a great expansion of trade and the rise of capitalism. But Locke's concern with happiness and enlightened self-interest and prudence were part and parcel of his tentative, undogmatic, gradualist, anti-authoritarian approach to public affairs. To preach enlightened self-interest has in practice proved a wiser principle in politics than to seek converts to high-minded ideologies. Those who think happiness important are more likely to promote it than those with loftier aims and greater certainty of purpose. And prudence is quite a good prophylactic against dictatorship or uncritical acceptance of ideologies. Locke was not a visionary who sought to inspire us with ideals such as Plato's Republic or Marx's classless society. He was a champion of the open society, which many visionaries since have done their best to destroy.

The Enlightenment in Britain was therefore a period in which science and liberal ideas in politics were seen as inter-dependent. It was a period of optimism, not only because it was a time of liberation for the human spirit, but also because science and the development of trade offered the prospect of material prosperity. Locke and his followers thought that their kind of politics would lead to greater freedom and tolerance and I believe their hopes have proved justified. Tom Paine's *Rights of man*, another product of the Enlightenment, remains an inspiration and the founders of one of the world's great democracies, the United States, acknowledged their debt to the leading figures of the British Enlightenment. Thomas Jefferson declared that his three greatest heroes were Isaac Newton, Francis Bacon, and John Locke.

In more recent times, the Enlightenment has been attacked by postmodernists, among others, as the source of many of the evils they perceive in contemporary society. They blame it because the advance of science and technology that followed enabled Western Europe to colonize the undeveloped world and to impose its own cultural values on others. Some, including Isaiah Berlin, have criticized the *philosophes* for their

'faith in universal, objective truths in matters of conduct, in the possibility of a perfect and harmonious society, wholly free from conflict or injustice or oppression—a goal for which no sacrifice can be too great—... an ideal for which more human beings have sacrificed themselves in our time than, perhaps, for any other cause in human history'.[16]

It is true that some of the *philosophes* did believe that scientific principles could be applied to human conduct and had a Utopian vision. To that extent, Marxists were their intellectual successors and duly exacted their toll of human suffering. Many Enlightenment thinkers also believed in the inevitability of progress. But this is where a distinction should be drawn between the rationalist tradition, with its emphasis on the primacy of deduction and mathematics, a belief in certainties, and a willingness to change society fundamentally; and the pragmatic tradition of Locke and his British successors, whose approach was tentative and experimental and who eschewed certainties. Locke helped to create a mode of thought in Britain—practical rather than idealistic, gradualist rather than Utopian, and based on evidence and common sense rather than strict logic—which may not always have satisfied or inspired the deepest thinkers and the greatest minds, but which helped to save Britain from the revolutions and more extreme political movements that have afflicted some countries on the mainland of Europe. It has been said that in Britain practice dictated theory, whereas in many other states theory dictated practice.

Since the Enlightenment, liberal democracies have been established in Europe as well as America and in many other parts of the world, while science has transformed our lives. It is true that the course of progress has not been a story of steady, uninterrupted improvement without major setbacks on the way. The industrial revolution exacted a heavy toll in human suffering and the lives of those who worked in the 'dark Satanic mills' in its early days were no improvement on their former rural existence. Furthermore, there have been wars in which technology enabled the combatants to kill each other with devastating efficiency. Nevertheless, science made it possible to achieve a major qualitative change in the nature of society. In a pre-scientific agrarian society, life was a perpetual struggle in which the scarcity of resources almost inevitably led to a hierarchical, authoritarian form of organization.

Most people suffered or starved in accordance with ascribed rank. The well-known hymn *All things bright and beautiful* gives us the picture:

> The rich man in his castle,
> The poor man at his gate,
> God made them, high or lowly,
> And order'd their estate.

Science and technology have given people a chance to live a fuller life. Moreover, scientific learning, unlike previous forms of learning that sometimes seemed to go round in circles, built on itself, and new scientists, as Newton proclaimed, stood on the shoulders of their predecessors. Science therefore generated a sense of optimism about the future and was naturally cast as the engine of progress, associated with the gradual spread of political freedom as well as material wealth.

The decline of optimism about science

Today we are healthier than we used to be and live longer, thanks to modern medicine. We are better fed, thanks to modern agriculture. We can travel more widely and more safely than previous generations, thanks to technological advances in transport. Knowledge and education are more widely dispersed than ever before. On the face of it, science and technology have been hugely successful. Why then do we want to bite the hand that (literally) feeds us and question the benefits of science? What are the deeper reasons that have caused the optimism about science that marked the Enlightenment and later generations to fade and turn into widespread pessimism and suspicion?

First, of course, we must ask whether people do in fact feel suspicious about science and if so, how deep this suspicion goes. Opinion polls do not give a clear picture of what the public in the UK actually thinks today.[17] Some answers suggest it has not in fact lost faith in science at all. People still believe that generally science makes the world a better place. This belief seems to be supported by the enthusiasm shown for such new technologies as computers and mobile phones and by the huge sales of books about science. Stephen Hawking's *A brief history of time: from the big bang to black holes* was on the best-seller list in Britain for over two years, although it was notoriously more often bought than read beyond its first chapter. Again, when asked who people trust, doctors nearly always come top of the league and scientists somewhere in the middle. On issues such as pollution, nuclear power or BSE, people trust university scientists—but not government scientists or scientists in the pay of industry. Since scientists in universities either get money from government or industry, the category of trusted scientists rather fades away.

More particular findings, however, show attitudes of mistrust and pessimism, especially when it comes to issues that affect people most directly, such as food, health, and the environment. When it comes to environmental matters, those whom the public trusts most are pressure groups like Greenpeace and Friends of the Earth. Yet these are the very bodies that do most to arouse public concerns, that generally find themselves in conflict with independent scientific opinion, and, far from being objective commentators, have their own vested interest in spreading scare stories because they boost membership. It seems the public prefers to believe the bringers of bad news rather than good news and trusts lay commentators more than experts. Other poll findings show that the public does not know, or even care, what scientific evidence is and does not understand how the scientific method works, or the importance of peer review in establishing the reliability of research results. People turn in ever larger numbers to practitioners of alternative medicine. Scare stories about the measles, mumps, and rubella (MMR) vaccine gain widespread credence, although the vast majority of scientists and doctors assure the public that there is no evidence of a link with autism. Seventy-five per cent of people in the United Kingdom worry greatly about toxic substances, such as pesticides, in food, although food has never been safer or more carefully tested. A majority believes (contrary to the evidence) that GM food is dangerous to health and that organic products are safer and more environmentally friendly. They also believe that astrology is scientifically based. Our culture has become risk-averse and there is a widespread demand that new products and technologies should be positively proved absolutely safe before they are let loose upon the public, a requirement that would stop all innovation. Furthermore, a highly articulate section of opinion, strongly represented by comment in the press, expresses a profound malaise about the direction in which science is going and wants the public to have more control over what science does.

There are therefore good grounds for saying that there has been a change of mood from optimism towards pessimism, and so the question arises: what has caused the change? The answer is inevitably a matter of subjective judgment. I offer a number of explanations: the advent of nuclear power; an increasing concern about the impact of science and technology on the environment and a rise in the influence of environmental pressure groups; and finally a feeling that science is out of control because of the speed of change and because the techniques of molecular biology now appear to give it unlimited power over nature.

Fear of nuclear power

Technology had already provided ample evidence in the first half of the last century that it could be used for massively destructive purposes. In two world wars its products killed people on an unprecedented scale. Then came the atom bomb, followed by the hydrogen bomb, weapons that threatened to destroy the whole planet. The chances of all-out nuclear war may have receded, although so-called 'weapons of mass destruction' still have the potential to cause carnage on a large scale. But the fact that science had produced a technology that could turn the whole earth into a desert changed perceptions about technology from something ultimately beneficial into something that could do more harm than good.

Nuclear power not only shook public confidence in the benign nature of science, but also made people more suspicious about scientists. The dangers from nuclear tests, for example, were not explained at the time, largely because they were not anticipated. But when they did become more generally known, in the 1950s and 1960s, there was a widespread public feeling that they had been deliberately concealed. The *New Yorker* journalist Paul Brodeur, who later became the principal protagonist of the view that electro-magnetic radiation from overhead power lines causes cancer (a view contradicted by extensive epidemiological studies) originally made his name by claiming to have exposed a series of Cold War conspiracies. The military-industrial complex was combining with scientists, he argued, to foist a nuclear future on unsuspecting citizens. Certainly, official attitudes at this time gave good grounds for concern by Brodeur and others. Some strange figures seemed to exercise a sinister influence behind the scenes. Herman Kahn, head of the influential Rand Corporation in America, later reputed to be the model for the eponymous hero of Stanley Kubrick's satirical anti-war film *Dr Strangelove*, wrote two books, *On thermonuclear war* and *Thinking the unthinkable* (namely, about nuclear war), in which he seemed almost to relish the contemplation of Armageddon. (In the early 1960s, I heard Kahn say at a dinner party, without appearing in the least perturbed by the prospect, that nuclear war was more probable than not. He also, incidentally, denounced Britain for its lazy work practices and, when someone mentioned that current opinion polls showed that Britons were generally more contented than Americans, he remarked: 'It's a very bad thing for people to feel happy'.)

Even the use of nuclear energy for peaceful purposes, which initially promised to provide cheap energy as a benefit to offset the threat from nuclear weapons, has come to be seen as a threat to safety. It was to be the fuel so cheap 'it wouldn't pay to meter it'. Governments promoted it with fervour and minimized its risks; by contrast, the anti-nuclear movement maximized them. The antis won the battle for public opinion in the UK. When there was an accident in 1979 in a nuclear reactor at Three Mile Island (Harrisburg) in the United States, although nobody was killed, public confidence in nuclear energy was shaken throughout the world. A disastrous accident at Chernobyl in the Ukraine in 1986 did kill people and left others worried that its full damage might not become apparent for generations. Although their fears proved greatly exaggerated, the future prospects of the nuclear industry as a whole were severely damaged. These events and the debate about nuclear power that followed have also had three other consequences: firstly, scientists became part of the political debate; secondly, since they appeared on both sides of the argument, the feeling grew that even the scientists themselves did not know what would be the effects of their innovations; thirdly, suspicion increased that there was a conspiracy between government, scientists, and industry to pursue technological progress regardless of social consequences and risks. Other major industrial accidents, such as an explosion in a chemical plant in Seveso in Italy in 1976 (an incident famously described as 'the poison that fell out of the sky', but which fortunately had no immediately fatal results), and the escape of poisonous gas from a chemical plant at Bhopal in India in 1984, which is estimated to have killed over 3000 people, conveyed the message that modern technology was making the world a more dangerous place to live in and that governments and scientists were trying to cover up the risks.

The rise of the environmental movement

A second, and perhaps the dominating, factor has been the rise in public concern for the environment. The environmental movement has a long lineage dating back at least to the latter part of the nineteenth century. The German biologist Ernst Haeckel appears to have been the first to use the term *Ökologie*, to describe the science of the relations between organisms and their environment, and he imprinted some of its present characteristics on the ecological movement from the start: a strong ethical and political content and a refusal to take an anthropocentric view of the world. Anna Bramwell, who has traced the history of the ecological movement, stresses its evangelical nature, which was evident long before the crusaders of Greenpeace and Friends of the Earth took up the cause.[18] The movement had particularly strong roots in Germany. The Romantic notion of a mystical union between a people and its homeland—which was later to find expression in the Nazi concept of *Blut und Boden* ('blood and soil')—had long conditioned German thinking about nature. Rudolf Steiner was another nature devotee who had a profound influence on his contemporaries, and was indeed one of the founding spirits of the Soil Association.[19] The philosopher Martin Heidegger has been described as the metaphysician of ecologism[20] and declared that 'man should be the shepherd of the earth'. But the call for a way of life more in tune with nature also had its influential supporters in England, including D. H. Lawrence, Rolf Gardiner (one of the founders of the Soil Association), Henry Williamson (author of *Tarka the otter*) who was often regarded as the voice of ecology in England, and two prominent Catholic intellectuals, G. K. Chesterton and Hilaire Belloc, who advocated a policy of 'back-to-the-land' and looked back to a glorified past in the Middle Ages. At this stage the ecological movement had strong right-wing connections: Gardiner and Williamson both expressed sympathy for the Nazis, Rudolf Steiner joined the Nazi party in its early days, and Heidegger was also an active and vociferous party member.

In the period after the Second World War, the environmentalist movement became associated with the left rather than the right. Nowadays it has a profound impact on policy, particularly in the life sciences, in almost every European country. Campaigns against genetically modified crops, hormone-treated beef, experiments using animals, and greenhouse gases are all more strongly supported and more influential in Europe than in the USA. Green parties are represented in some strength in many European Parliaments. Yet the origins of modern green politics lie firmly in the USA. The Green pressure group Friends of the Earth was founded there. Rachel Carson, an American, has the strongest claim to the title of mother of the modern environmental movement, while the US social rights activist Ralph Nader may be called the father of the consumer movement, at least in its highly politicized form.

I would rate Rachel Carson's book *The silent spring*, published in 1962, as one of the most influential books of the last century. It won converts by the tens of thousands and certainly had a profound effect on me. She conjured up a most eloquent doomsday scenario of the consequences of indiscriminate use of insecticides: a landscape in which no flowers bloomed, no birds sang, and the rivers were devoid of fish. She told the fable of a town in the heart of America in which mysterious maladies swept through flocks and doctors were faced with sudden and unexpected deaths among their patients: 'Everywhere there was a shadow of death'. The principal culprit was the insecticide DDT, and her warnings of doom struck a particular chord because she claimed that bird populations diminished by DDT included the national symbol, the American bald eagle.

Rachel Carson performed a service in alerting people to the danger to the environment of indiscriminate spraying of insecticides such as DDT. But she greatly overstated her case. While it is widely believed that predatory bird populations were affected by the use of DDT, it seems that the bald eagle was not one of them. It was estimated that their number in the United States was about 200 in 1941, six years before the widespread use of DDT, and some 900 in 1960.[21] She was also wrong about the effect of DDT on migrating ospreys in the United States and on peregrine falcons in Britain. The numbers of ospreys observed at Hawk Mountain, Pennsylvania, increased from 191 in 1946 to 630 in 1972. As for peregrine falcons, the British government's Advisory Committee on Toxic Chemicals, in its review of organochlorine pesticides in Britain in 1969, concluded: 'There is no close relation between the decline in population of predatory birds, particularly the peregrine falcon and the sparrowhawk, and the use of DDT'.[22] What is more, laboratory experiments failed to establish the link between DDT and eggshell thinning that Carson claimed.[23]

Moreover, Carson forecast that DDT would cause cancer in human beings, an allegation that was also made by the World Wildlife Fund when it campaigned for a total ban on DDT (a campaign that it has now modified), and she claimed there was a connection between DDT and the rise in the number of cases of hepatitis. 'For the first time in human history', she wrote, 'every human being is being exposed to contact with dangerous chemicals from conception until death'.[24] Allegations about cancer and hepatitis have proved groundless. The conclusion of the National Academy of Sciences in the United States in a report to the Environmental Protection Agency was: 'The chronic toxicity studies on DDT have provided no indication that the insecticide is unsafe for humans'.[25] (In chapter 7 I refer to the hugely beneficial effects of DDT.)

From its earliest days, the environmental movement was inclined to regard science and technology not as allies but as enemies. Many of the early ecologists looked back with nostalgia to the supposed harmony between nature and society in the Middle Ages and regarded the birth of science as the start of 'a mechanistic, rapacious, inorganic attitude towards nature'.[26] Carson saw science and technology as dangerous because they were part of mankind's mistaken attempt to control nature. 'The control of nature is a phrase conceived in arrogance, born of the Neanderthal age of biology and society, when it was supposed that nature exists for the convenience of man'.[27] Pesticides were 'the elixirs of death', the consequence of the zealous pursuit of modernization. Or, as Sir Julian Huxley put it in his introduction to the British publication of *The silent spring*, 'The present campaign for mass chemical control, besides being fostered by the profit motive, is another example of our exaggerated technological and quantitative approach.' From its birth, therefore, the environmental movement embraced all the basic elements that characterize eco-fundamentalism today: exaggerated claims about damage to the environment and health that are not supported by evidence; a rejection of modern science and technology because they seek to control nature, together with a call for a 'return to nature' instead and the identification of science and technology with capitalism and the profit motive.

Ralph Nader started his public career as the consumer's champion, whose main concern initially was to shake up the complacency of society by making consumers more aware of risk. His successful book *Unsafe at any speed*, published in 1965, blamed car accidents not on the behaviour of drivers but on the 'designed-in dangers' of the American automobile. Early on, he linked the new consumer movement with environmentalism and gave it a strong campaigning flavour. Regulation against risk, including environmental risk, was taken up enthusiastically by successive American administrations, mainly because it was popular, possibly in Lyndon Johnson's case because it was a diversion from preoccupations with Vietnam,[28] and perhaps also because larger corporations found that regulations requiring higher environmental and safety standards discouraged competition from smaller rivals, who found compliance with them extremely onerous. Predictably, more extensive regulation against risk has had the effect of making people more worried about risk and more distrustful of government and science. After all, the thinking goes, if they pass all these laws, there must be something to worry about.

One of the early successes of the environmental movement in the United States concerned events at Love Canal in 1978, which are discussed in more detail in chapter 7. Residents in homes built on land that had been previously used for waste disposal claimed that the leakage of toxic waste into the canal had caused birth defects in their children. Their campaign led to the declaration of a national emergency and the relocation of the residents to a safer place. It was regarded as a triumph for local activists and people-power, and had a profound effect on environmentalist activity. An anti-toxic movement was created and mushroomed: there were 242 active groups in the United States in 1981, 5000 by 1989.[29] This movement accused conventional science of having an inbuilt bias against looking for adverse impacts on health. Activists rejected a 'rationalist, probabilistic approach'[30]: in other words, science should yield to the judgment of lay people, a theme that is echoed in much criticism of science today. Ralph Nader duly announced that America had entered the 'carcinogenic century', *Time* magazine announced 'The poisoning of America', and *Newsweek* described how our insatiable desire for consumer goods threatened us with destruction through the by-products of technology.

Europe then took up the baton and ran even faster. The European Union declared a moratorium on the commercial cultivation of genetically modified crops, officially promoted so-called pesticide-free organic farming, and has enshrined the Precautionary Principle in its treaties. At least part of the reason for the greater disquiet in Europe about science and some of its applications is a series of health scares and actual disasters that have profoundly affected public opinion, especially in Britain. Two events in particular had a deep impact on public attitudes to science and to experts generally. The first was the unfortunate history of the sedative drug thalidomide. Its approval in Europe in the late 1950s was followed by the birth of a large number of malformed children and the memory of the disaster still lingers. The second was the saga of bovine spongiform encephalopathy (BSE) in the 1980s and 1990s, when government spokesmen assured the public that eating beef was safe. Whenever authority is questioned, reference is made to the traumatic experience of BSE as conclusive evidence that experts cannot be trusted because they make mistakes and that assurances by authority cannot be trusted because they hide the truth.

It is worth observing how the consumer movement in Europe has changed over the years, following in Nader's footsteps. When the Consumer Association was founded in Britain in the 1950s, it provided an objective consumers' guide to goods and services. It was determined to remain strictly independent of all outside influences and to be non-commercial and non-political. The evidence-based approach was its guiding light: how did goods stand up to rigorous independent testing by experts? Now the Consumer Association in Britain, like its international counterpart and other consumer organizations, is a political, campaigning NGO, which has allied itself with other NGOs in calling for a further moratorium on the commercial growing of GM crops and seems to me to attach more importance to 'ethical standards' than the need to find out actual facts.

The fear that science is out of control

A third factor that has undermined belief in the benevolence of science is a common fear that scientists are determined to rush ahead wherever new discoveries carry them, irrespective of social consequences. This fear is now most frequently expressed in relation to the life sciences. Indeed, at one stage, in 1974, scientists themselves were sufficiently worried that biological research was leading them into dangerous territory that in an unprecedented move they agreed to a moratorium on research into recombinant DNA until they had fully considered the risks and how to control them. At the famous conference at Asilomar in California in 1975 they decided there were in fact no good reasons for banning work on DNA, provided there were adequate safeguards. But if scientists themselves had qualms, is it surprising the public is worried? Indeed, the argument that science could go too far cannot be lightly dismissed.

We can now make domestic animals bigger and leaner and do so faster and more efficiently without recourse to traditional breeding. There is no technical reason why farmed animals might not be genetically modified to become double their present size. This doubtful success has already been achieved with carp and salmon. Why not produce elephant-sized cows? We have already created 'geeps', half-goat and half-sheep. There is, it seems, no limit to what we can do to our fellow creatures if we put our clever scientific minds to it, stocking our farmyards with monsters of all shapes and sizes, though presumably we would only do so if they served some useful purpose to mankind.

At least as disturbing is the perceived threat of a revival of the eugenics movement. We may soon be able to eliminate genetically transmitted diseases by replacing defective genes with healthy ones. So far, progress in modifying genes responsible for such incurable diseases as Huntington's chorea or cystic fibrosis has been slower than expected, but success will no doubt eventually come. In time, we may be able to apply gene therapy to the reproductive cells themselves. Will we perhaps be able to turn the sickly into the healthy, the weak into the strong, the small into the tall, the ugly into the beautiful and perhaps even the stupid into the clever? James Watson, who shared the Nobel prize for his part in discovering the structure of DNA in 1953, probably the greatest discovery of the last half-century, has stated that he would regard such developments as desirable. Again, we have cloned Dolly the sheep, although she died prematurely and was not ultimately in the most robust of health. Some doctors have already claimed that they can, and even that they would, clone human beings, although these claims were shown to be unfounded. But supposing they can, will some of the rich not find such offers irresistible? Perhaps the Brave New World is just around the corner after all. Perhaps it is not surprising that some people believe that science is too dangerous to be left to scientists and that the public must have more say in the future of research.

What is often forgotten is that most of these moral dilemmas are not new, and in some cases what can now be achieved by altering genes can already be achieved by other means. As Ronald Dworkin has pointed out,

What is the difference between inventing penicillin and using engineered and cloned genes to cure even more terrifying diseases than penicillin cures? What is the difference between setting your child strenuous exercises to reduce his weight or increase his strength, and altering his genes, while still an embryo, with the same end in view?[31]

Eugenics can be practised effectively without the need for modern advances in genetic science, as the ancient Spartans proved. Eugenics were also much in vogue in the 1930s in the USA, when several states enacted laws for the compulsory sterilization of mental defectives and criminals, and were notoriously applied by the Nazis. What stops the practice of eugenics is the revulsion of society, and attitudes will vary in different societies and cultures at different times. In my view, the way research into stem cells from human embryos has been regulated in the United Kingdom is a model of rational enlightenment. It allows research that may lead to cures for conditions such as Parkinson's disease and diabetes, but explicitly outlaws the socially unacceptable development of cloned human beings. Other countries, largely for reasons of religious conviction that the fertilized ovum must be regarded as a human being, prohibit such research. As in many other fields, the laws of a country can effectively outlaw practices that society finds repugnant.

Of course it is true that biotechnology (like anything else) offers opportunities for abuse. That has been true of virtually every major technological advance in history. Ever since Prometheus, the patron of discovery, gave mankind the gift of fire, we have played with fire. What is different today is that a climate of pessimism and risk aversion has developed so that many people seem to wish that Prometheus had never bestowed his gift—which would certainly have saved him a measure of discomfort on his Caucasian rock. Fear about the direction in which science is taking us leads many people to hanker for the past, for the days before the birth of modern science when we were closer to nature and trusted in nature, because 'nature knows best'. Hence the extraordinary popularity of homeopathy and various alternative forms of medical treatment that have no scientific basis and the dramatic growth of organic farming.

2

Medicine and Magic

In days of old, when men lay sorely tried
The doctors gave them physic, and they died.
But here's a happier age, for now we know
Both how to make men sick, and keep them so.

Hilaire Belloc

IT could be said that the ancient Egyptians were the first scientific doctors.[1] Granted, their views of anatomy were eccentric, since they believed that the heart was the centre of thought, the liver produced the blood, and the brain cooled it. Egyptian physicians were fatalistic about their treatment of illness, since they believed that the gods ultimately determined the fate of their patients. They also believed in magic. But, crucially, they wrote down what they did, recorded their experiences, and taught other doctors. The Greeks built on the lessons of the Egyptians. The body of medical writing attached to the name of Hippocrates was not only famous for its medical ethics, but also for its belief that diseases were not cured by magic, had rational causes, and needed rational cures. For example, Hippocrates taught that the so-called 'sacred' disease, epilepsy, had nothing to do with the gods, but was caused by phlegm blocking the airways and the convulsions were the body's attempts to clear the blockage.[2] Not that the cures prescribed by the Hippocratic school were necessarily helpful, but they took the enlightened view that what people needed was a healthy diet, exercise (or sometimes relaxation), and fresh air. Galen, a Greek who lived in the second century AD, treated wounded gladiators and learnt a great deal about the human body. He discovered that the brain was the centre of thought, that the nervous system controlled movement, and that blood moved throughout the body. While no one in his day was allowed to dissect human corpses, he studied the anatomy of pigs, and optimistically and sensibly assumed that human bodies were much the same. The Romans, following in the steps of Hippocrates, promoted public health by building viaducts to provide towns with fresh water and clean baths and by constructing sewage works. Some useful steps towards the advancement of medicine were therefore taken in the ancient world.

After the fall of Rome, progress towards a more scientific medicine came from the Persians and the Arabs. Persian doctors continued the Greek practice of cataloguing diseases and their treatments. Razes, in the tenth century, noted the difference between measles and smallpox. Avicenna (Ibn Sina), in the eleventh century, found that dirt infected wounds. The Muslim world was noted not only for the skill of its doctors but also for its enlightened approach to the administration of medicine: by the twelfth century, Baghdad, centre of the Abbasid caliphate, had a public health service that provided 60 hospitals in which the service was free. Hospitals throughout the Muslim world kept separate wards for different diseases and were regularly inspected. Nothing comparable existed in non-Muslim Europe, which during the Dark Ages and medieval period turned back from the idea that diseases had rational causes and explained them instead in terms of divine displeasure or magic spells. In the Christian world, prayer was regarded as the principal remedy, addressed to the patron saint of each disease. Saint Lucy, for example, who was said to be very beautiful and who, according to one legend, showed such devotion to virtue that she plucked out her eyes to make herself less attractive to admiring males, became the patron saint of eye diseases. Saint Blaise became the patron saint of throat diseases because he saved a child from choking on a fishbone. Herbs were much relied on, as they still are by devotees of alternative medicine (and indeed, in their modified form by scientific medicine). Many herbs were associated with a local saint and were prescribed in certain seasons, during certain stellar alignments, for patients of particular zodiac signs who were showing a certain characteristic of one of the 'humours' (four bodily fluids—blood, phlegm, yellow bile or choler, and black bile or melancholy—which were thought to determine a person's physical and emotional state). The idea of the humours, one of the less scientific concepts adopted by the Greeks from China and India, continued to have a profound influence throughout the world for many centuries. Indeed, the doctrine survives in some alternative medical treatments to this day.

The early history of medicine is not just of academic interest. The supporters of many forms of alternative medicine today claim it as a virtue that they are based on practices which have been in use for centuries, as if it was some sort of recommendation that they date back to medieval times or even earlier. A more accurate description might be that they are steeped in ancient superstition. It also seems somewhat perverse to treat medical practice as some kind of antique furniture whose value increases with age. The Middle Ages were not a golden period of history, in which happy peasants enjoyed a blissful natural existence. Instead, life was mostly nasty, brutish, and short, not least because of the ravages caused by unpleasant diseases which modern medicine is now able to prevent or treat. However, there has always been a struggle between the critical, analytical approach of reason and the instinctive disposition of human beings to believe in magic. Medicine in mediaeval times and the ancient world relied heavily on witchcraft, and even in modern times it has never managed to rid itself entirely of an element of magic and ritual. That it sometimes appeared to work (otherwise the shaman, witch doctors, and physicians would not have stayed in business) owes something to the placebo effect. Indeed, it has been claimed that the 15 remedies described in the oldest Sumerian medical tablet, dating back to 2100 BC were placebos, as were all the remedies of the ancient cultures.[3]

Whereas modern science developed rapidly after Bacon and the Enlightenment, medicine lagged behind. Doctors continued to espouse the concept of the bodily humours and to administer purges and practise bloodletting without any evidence to support either concept or treatment. A number of poisonous herbs and compounds were used to ensure that the humours in the body were kept in balance.[4] For a long time, purges were a way of medical life in the Western world. They were still in vogue, both in Europe and in the United States, until the end of the nineteenth century.

However, by the end of the twentieth century medicine had been transformed. Initially public health reforms were the most important. These reforms, later accompanied by the advent of antibiotics, a host of life-saving drugs, vaccines against life-threatening epidemics, as well as better diet and safer food, have extended life-expectancy in the developed world in a way that could not have been predicted when the century began. In the developing world too, average life expectancy has increased dramatically in the last fifty years, from 41 years in 1950 to 65 in 1998.[5] Altogether evidence-based medicine has made a huge contribution to our physical well-being. It can justly be described as one of the greatest achievements of mankind.

Yet, curiously, this progress has not led to greater public confidence in the state of our health or to a wider trust in medical science. Not only are there constant health scares, most of them unjustified, but alternative treatments that are not science-based are highly popular and are a feature of almost every advanced industrial society. Perhaps it is part of a sense of malaise that we have lost touch with nature because our union with nature has been disturbed by science and technology. There is a widespread, instinctive feeling that 'nature knows best'.

Indeed, many medically qualified doctors practise homeopathy and offer alternative medicine as part of their services. Most doctors are not scientists and many are not as committed to the evidence-based approach as one might expect from a profession whose success today depends so extensively on the progress of science.

Alternative medicine

It is important to clarify what is meant by alternative medicine, especially as there can be some overlap with 'orthodox' or 'conventional' medicine. Alternative medicine, part of which may also be described as complementary medicine, covers a broad spectrum of approaches. It ranges from osteopathy, chiropractice, and acupuncture, to homeopathy, herbal treatments, reflexology, aromatherapy and includes even more mystical practices such as Ayurveda.[6]

Some manipulative treatments, such as chiropractice and osteopathy, are not always clearly distinguishable from their conventional counterpart, physiotherapy; furthermore, they are properly regulated in the UK, unlike other forms of alternative medicine.[7] Many drugs initially derived from plants have been properly tested and proved to be effective and are now part of orthodox medicine. Aspirin (a synthesized form of salicylic acid, originally derived from willow bark) and quinine are obvious examples, as is digitalis from the foxglove, used in the treatment of heart conditions. Some herbal extracts that have been used for centuries on the basis of traditional knowledge have been vindicated as the active agents were discovered and we found out how they worked.[8] Even old wives' tales have sometimes turned out to be based on accurate observation, as Edward Jenner discovered when he investigated the widespread belief that milkmaids did not suffer from smallpox and found that they had become immune to it as a result of their exposure to cowpox.

By contrast, there are other practices that make no claim to any scientific basis. Ayurveda, for example, is based on the concept that the body is defined by three 'irreducible physiological principles' called *vata, pitta*, and *kapha*,[9] which in some ways resemble the old European notion of the four humours and the Taoist principles of *yin* and *yang*, and which have to be kept balanced for harmony. Aromatherapy is based on the healing properties of essential oils, which are supposed to represent the spirits and souls of the plants from which they have been extracted.[10] One form of treatment, acupuncture, inhabits a sort of twilight zone: sometimes shown to be effective, sometimes almost certainly dependent on the placebo effect.[11] However, as Ray Tallis has pointed out, 'acupuncturists require one to believe ideas about illness, for which there is no evidence, other than the sacred texts of Chinese medicine: that there are patterns of energy flow (Qi) throughout the body that are essential for health; that disease is due to disruptions to this flow; and that acupuncture corrects the disruptions'.[12]

Some practices do no harm and are popular because they provide people with an agreeable experience: aromatherapy (which simply means absorbing pleasant smells) and Indian head massage are so enjoyable in themselves that it hardly matters whether they solve any medical problems. The physicist, Richard Feynman, in his essay *Cargo cult science*, describes his experience of reflexology, in which in a bath on the warm Californian coast he had his feet stroked by a rather pretty, somewhat skimpily dressed, and pneumatic young lady. He writes that he did not feel very motivated to question the evidential basis for such a pleasant experience.[13] On the other hand, Ayurveda, a form of healing linked with a mystical approach to nature, may involve serious risks to health: in his book *Bad medicine*, Christopher Wanjek describes the standard Ayurvedic remedy for preventing and reversing cataracts as 'brush your teeth and scrape your tongue, spit into a cup of water and wash your eyes with this mixture'. He cites a report in the *American Journal of Medicine* that Ayurvedic treatments can cause hallucinations, anxiety, depression, insomnia, and gastro-intestinal problems.[14]

The growth of alternative medicine has been impressive. In the United States there are some 425 million visits a year to practitioners of alternative medicine, compared with 388 million visits to primary care physicians.[15] In Germany, homeopathy is now considered part of mainstream medicine. In Britain, 40 per cent of general practices provide some form of alternative medicine.[16] Many doctors not only dispense homeopathic remedies alongside conventional drugs, but have themselves become qualified homeopaths. Almost every chemist shop now has a section devoted to the display of complementary preparations.

It is clear that alternative medicine is gradually becoming respectable. The *British Medical Journal*, in a recent issue, advocated 'integrative' or 'integrated' medicine to combine orthodox and alternative medicine, and invited the Prince of Wales to contribute an article, although he is better known for his rejection of modern science in favour of mysticism and for his belief in a return to nature than for his medical expertise.[17] A recent guide to hospitals providing the best care for breast cancer in a reputable Sunday newspaper cited the availability of aromatherapy, reflexology, homeopathy, and acupuncture as among its criteria of excellence.[18] Some medical schools now include alternative medicine as part of the curriculum, although still only in the form of special study modules in which students learn about patients' beliefs and the alternative treatments that patients ask for.[19] In the United States, however, 75 out of 125 medical schools have no qualms about offering education in alternative therapy and giving advice on prescribing complementary preparations.[20] It is as if the teaching of astrology or pre-Copernican astronomy were made part of the national schools curriculum. (In some states in America, of course, some schools now teach creationism—or 'creation science' as its proponents would have it—alongside evolutionary science.)

In principle, it should be possible to carry out randomized control trials to establish the effectiveness of alternative medicine as is done for all other forms of medical treatment, but in practice not many such trials have been performed. There are several reasons why this is so. One is that practitioners often take the view: 'What's the point? We know it works.' Less convincingly, some practitioners argue that their treatment is so effective that it would be unethical to submit it to randomized trials. Cynics might point out that if a doctor practising conventional medicine finds that a pet theory is disproved, this does not invalidate all conventional medicine. If, on the other hand, the effectiveness of a particular form of alternative medicine in which a particular practitioner specializes is disproved, the whole justification for the practice and the validity of the practitioner's training is brought into question. Anxiety that the results of trials might not be favourable is one of the reasons for the paucity of trials.[21] Furthermore, what research has been done has often been of poor quality, for the reason that not many of those who practise alternative medicine are experienced at conducting research. An even more sceptical view is taken by Professor Richard Dawkins, who has defined alternative medicine as 'that set of practices which cannot be tested, refuse to be tested, or consistently fail tests'.[22]

It is highly significant that practitioners of alternative medicine themselves acknowledge that scientific research into its effectiveness may be 'inappropriate'. The editor of the American journal *Alternative Therapies* has admitted this quite openly. He has argued that

many alternative interventions are unlike drugs and surgical procedures. Their action is affected by factors that cannot be specified, quantified, and controlled in double-blind designs. Everything that counts cannot be counted. To subject alternative therapies to sterile, impersonal double-blind conditions strips them of intrinsic qualities that are part of their power. New forms of evaluation will have to be developed if alternative therapies are to be fairly assessed.[23]

What he is really saying is that we have no way of testing the effect of most alternative medicines because they have intrinsic qualities that we just have to take for granted. Randomized double-blind trials are important because they eliminate any bias that might be introduced by either the patient or the doctor knowing whether the patient has been assigned to the test group or the control group. Richard Dawkins' definition is confirmed.

Homeopathy

Homeopathy can perhaps be regarded as the *pons asinorum* of attitudes to complementary medicine. (The *pons asinorum*—literally, a bridge asses have to cross—is Euclid's fifth proposition, a sort of intelligence test for beginners.) Homeopathy is widely believed to

be effective, and has often been claimed to have scientific justification. This claim, to put it mildly, is short of proof or plausibility.
Homeopathy is based on the law proclaimed by the German physician Samuel Hahnemann in 1796 that *similia similibus curantur*,
or 'like cures like'. He derived his law from a single experience:
trying to find out why quinine relieved the symptoms of malaria,
he took some quinine and then developed chills and a fever.
From this, by a leap of faith that seemingly required no further
evidence, he concluded that substances that produce a particular
set of symptoms in a healthy person can cure those same symptoms displayed in someone who is ill. He then proceeded to test
various natural substances to find out what symptoms they produced and prescribed them for illnesses that produced those same
symptoms.

However, many natural substances, including quinine, are toxic.
To avoid the inevitable effect of administering a poison, he diluted
his medications in water and claimed that the more dilute they
were, the more effective they became. This led him to pronounce a
second law, the 'law of infinitesimals', that the more a medication
is diluted, the more patients will benefit. Although continued dilution means that ultimately not one molecule of the original substance will be present, homeopaths claim that it has left an imprint
on the water, which 'remembers' the substance after it has been
diluted away. A dilution of one to the power of 30 is fairly standard, which means that the original medicine or herb has been
diluted to one part in 1,000,000,000,000,000,000,000,000,000,000.
As Robert Park observed, this means that to swallow one molecule
of the actual compound, one would have to drink 7874 gallons of
the appropriate homeopathic solution.[24] Some substances are
diluted by 10 to the power of 400, that is, one part in 1 followed by
400 zeros.

The first of Hahnemann's laws of homeopathy, treating like
with like, is not supported by any evidence whatsoever. It is no
more scientific than the medical practice prevalent at the time he
made his 'discovery', of suppressing symptoms by a treatment of
opposites. His second law of infinitesimals depends on pure magic,

or miracle. Producing a remedy by extreme dilution in water and some sort of 'memory' effect on the water, is like the miracle of turning water into wine but without the benefit of divine intervention. It is claimed that some homeopathic remedies have been shown to have an effect; but unless one believes in magic or miracles, this must be due to the placebo effect. Homeopaths claim that they have done double-blind randomized trials which show significant differences between homeopathic medicines and placebos, but according to the Royal Society the evidence from these trials is 'insufficient for definitive conclusions because most trials were of low methodological quality'.[25]

Why, in spite of this, is homeopathy, and alternative medicine in general, so popular? There are several explanations. One is suspicion of pharmaceutical companies—many people believe that these powerful multinational companies are motivated by the need to make profits rather than curing the sick. Another is a wider suspicion of modern medicine as part of the belief that we should 'go back to nature.' Then there is the fact that in the UK doctors in the National Health Service are overworked and have little time to spend with their patients, whereas homeopaths may spend hours with them listening attentively to all their concerns. Tender loving care has always been an important part of good medical practice, for which nowadays not many doctors in the public service have time. Orthodox medicine often has to admit areas of uncertainty and, occasionally, the absence of any known cure. Alternative medicine admits no doubts and offers a cure for all conditions. Finally, the placebo effect can induce important beneficial effects in some people under certain circumstances.

At the end of the Second World War, for example, an American anaesthetist named Beecher gave battle casualties injections of saline when supplies of morphine had been exhausted. To his astonishment he found that these injections effectively relieved severe pain. Pain can be a subjective phenomenon and it should perhaps not be surprising that, if patients believe they are getting morphine when in fact they are receiving salt water, they feel less pain.

There are many surprises and misunderstandings about placebos. A study of their effects shows that placebo injections are more effective than placebo pills, that brand name placebos are better at relieving pain than generic ones, and that blue placebos are better placebos than red ones, except for Italian men.[26] Germans with ulcers respond twice as well to placebos as the rest of the world, but in the case of placebos to reduce blood pressure, Germans have the lowest response rate in the world. The answer to these apparent absurdities, according to the study, is that the crucial factor is the attitude and quality of the doctor. The more convinced the doctor is that the treatment will work, the more likely it is that it will, hence the need for trials to be 'double-blind', that is, doctors too must not know which treatment a patient is receiving. As George Bernard Shaw wrote in *Misalliance*: 'Optimistic lies have such an immense therapeutic value that a doctor who cannot tell them convincingly has mistaken his profession'.

Herbal medicines

In the case of several herbal extracts it has been claimed that they are in fact more effective than placebos. Some, such as digitalis, have been tested and vindicated. Others in common use have not. In some cases the evidence is ambivalent. For instance, it is widely accepted that preparations made from St John's wort (*Hypericum perforatum*) can help cure depression. It is described in Solomon's *Song of songs* as the 'Rose of Sharon' and was used by Paracelsus to treat nervousness, skin wounds, and abdominal pain. One study by the National Institute of Health and the National Center for Complementary and Alternative Medicine of its efficacy in the treatment of depression found 'no statistically significant difference between St John's wort and placebo'.[27] Another study found that in nine out of thirteen randomized clinical trials treatment with *Hypericum* was significantly better than placebo.[28] Unfortunately, the extracts used contained a number of active ingredients, and different amounts of them, so that combining the results of such studies of herbal preparations is problematic. However, experimental studies have also shown a favourable effect, and one of its constituents (hyperforin) is superior to placebo and as effective as orthodox antidepressants in the treatment of mild to moderate depression. It is remarkably safe, but there may be serious adverse interactions with a number of other drugs, including oral contraceptives and some conventional antidepressants. Since the one constituent is responsible for both therapeutic and adverse effects, safer analogues of hyperforin could be synthesized, an evidence-based approach that could help many patients. At present, extracts are marketed both in the UK and USA as a food, although 'the possible self-medication of St John's wort by patients suffering from major depression is of concern'.[29]

Arnica, a cream that many users believe helps to soothe the bumps and bruises suffered by children, has been one of the best-selling complementary medical preparations for many years. Yet a study carried out by Edzard Ernst, Professor of Complementary Medicine at Exeter University, showed that arnica has no effect, either in soothing pain or accelerating healing. It was declared to be a complete waste of money.[30] Again, echinacea is reputed to be effective in the treatment of colds. Yet recently a carefully conducted double-blind trial found no difference in effect between a group of people given echinacea and those given a placebo. The conclusion was clear: echinacea did not work.[31]

One reason why herbal medicines are popular is a belief that what is 'natural' must be safe. In fact many plant extracts are highly toxic, because plants have developed poisons to protect themselves from being eaten by animals. St John's wort, for example, can often cause several unpleasant side-effects, such as 'dry mouth, dizziness, gastrointestinal symptoms, increased sensitivity to sunlight and fatigue'.[32] A *British Medical Journal* publication devoted to complementary medicine states: 'Herbal medicine probably represents a greater risk of adverse effects than any other complementary therapy'.[33] Indeed, the danger of misuse is aggravated because herbal medicines are not regulated. A government move to impose regulation was withdrawn after representations from the herbal remedy industry.[34] Alternative medicine is big business and is supported by a powerful pressure group.

The disadvantages of alternative medicine

Overall, the current medical scene in the UK, and indeed elsewhere, presents a contrasting picture. On the one hand, medical knowledge continues to advance through scientific research, with every prospect that in time there will be cures for many of today's incurable illnesses, such as Parkinson's disease, Alzheimer's disease, and various forms of cancer. By contrast, the Back-to-Nature movement takes us back to the era before medicine was based on science and ignores what scientific medicine has already achieved. It may be argued that, despite the adverse side-effects of some herbal remedies, alternative medicine does relatively little harm and, through the placebo effect, some good, and it should not therefore be a matter for general concern. There are many thousands of people who are sincerely convinced of its value and who will refuse to accept the verdict of double-blind trials that efficacy is solely due to the placebo effect. They have derived great comfort from these forms of treatment. If people want to spend their money on treatments that have no rational basis but provide them with comfort, why not let them? We tolerate Christian Scientists, why not devotees of homeopathy and reflexology?

It is not the role of the state to intervene unless particular activities are shown to cause positive harm. 'Live and let live' is a basic principle of a civilized society. But there may still be some activities we should seek actively to discourage and in some cases to regulate for the effective protection of the public.

Firstly, mixing herbal remedies with common prescription drugs can actually endanger life. St John's wort, for example, which is widely used, can react dangerously with warfarin and a recent study discovered that one in five patients on warfarin were also taking complementary medicines and that nine out of ten of these had not informed their doctor[34]. Secondly, practitioners of alternative medicine may persuade people to forgo conventional medical treatment that they need. If the complaint is one that is not serious, or is one from which people are likely to recover anyway, no harm and possibly some good may flow from treatments that have no scientific basis but make people feel better. On the other hand, in the case of serious diseases such as cancer, ignoring conventional medicine could prove fatal. From time to time practitioners of alternative medicine are rightly found guilty of malpractice because they have failed to make a proper diagnosis or to provide proper treatment. The courts also intervene from time to time, justifiably, to stop parents denying their children medical treatment that could save their lives for religious reasons. It is true that at present most people who use complementary preparations still rely primarily on scientific medicine; but if the fashionable demand for alternative medicine continues to grow, the number of those who depend on it exclusively will also increase. When orthodox medicine is wholly rejected, as it is among some religious communities in America, the results can be devastating. For example, among the Faith Assembly religious sect in Indiana, peri-natal mortality is 92 times greater than in Indiana as a whole and maternal mortality is the same as it was a hundred years ago.[35]

Much the most depressing example of the damaging effect of rejecting modern or 'Western' medicine in favour of traditional remedies comes from South Africa. President Thabo Mbeki rejected the evidence that AIDS is a sexually transmitted disease caused by a virus, that its transmission could be prevented and its effect treated by anti-viral drugs. Doctors, scientists, and AIDS activists who disagreed were vilified and threatened; the campaign to promote anti-HIV drugs was denounced as 'an attempt to commit genocide against black people', akin to 'the biological warfare of the apartheid era'. It could only benefit pharmaceutical companies. By contrast, one of Mbeki's former supporters, the Anglican Archbishop of Capetown, described his AIDS policies themselves as a crime as serious as apartheid. South Africa's policy on AIDS is an extreme but telling example of the dangerous consequences to which the rejection of scientific medicine can lead.[36]

Thirdly, the practice of alternative medicine has a corrupting influence on doctors and on the profession generally, because it encourages a non-evidence-based approach. It is a throwback to the days of leeches and purges and pre-Enlightenment practices based on superstition and magic, when life expectancy was half what it is today. It substitutes anecdotal evidence for clinical trials. Alternative medicine is big business and it is a business largely based on myths. Products sold for which claims are made that are false or cannot be substantiated should be subject to regulation. Pharmaceutical drugs must undergo strict tests before they are allowed on to the market. There would be an outcry if they were not. Why should herbal products be immune from similar control? Yet the market power of the industry has been allowed to protect them from effective and necessary regulation.

The reaction against the MMR vaccine

The dangers of rejecting the evidence-based approach is further illustrated by periodical outbreaks of popular hysteria over health scares. For example, when the media gave wide coverage to an allegation that the triple vaccine against measles, mumps, and rubella (MMR) caused autism in children there was widespread panic among parents in the UK.[37] Before this scare, a highly articulate campaigning group already existed that objected to vaccination *per se*, as part of their general mistrust of modern medicine. (Most of this group were believers in homeopathic medicine.) Rumours then began to spread that a rise in the number of cases of autism was linked to the introduction of the triple vaccine. It is not surprising that parents who find that their child has become autistic shortly after vaccination should suspect a link between the two events. When they hear that other parents have suffered the same experience, suspicion easily hardens into certainty. There is also, it should be added, a more rational explanation why parents might refuse to vaccinate their children, in that every vaccination carries some risk, however small. (The risk is probably not so different from that of taking a baby out for a journey by car or on a flight in an aeroplane). At the same time, the diseases against which vaccines offer protection, such as diphtheria, whooping cough, tetanus, poliomyelitis, and measles, have virtually died out in the UK and the chance of an unvaccinated child becoming infected is now almost negligible. A mass vaccination programme is nevertheless needed to provide herd protection, to prevent the reappearance of these diseases; but while this is a good social reason for accepting MMR, it may not persuade parents concerned only with the welfare of their own child.

In 1998, a paediatrician at a leading London hospital, Dr Andrew Wakefield, published a paper based on observations made on 12 children who suffered from autism and bowel disease, in which he advanced the hypothesis that in eight of the cases autism might be causally related to the bowel disease, and further, that the bowel disease symptoms might in turn be due to the measles virus in the MMR vaccine. The anti-vaccination campaign group promptly joined forces with the parents of autistic children and an anti-MMR campaign took off. In fact, a Danish study tracing the history of 400,000 children found that those who were given the vaccine were no more likely to suffer from autism than children who were not. There were similar findings from studies in Finland and the UK covering over a decade and involving millions of children. No association was found between bowel disease itself and autism. Indeed it was pointed out that there was a much stronger association between the possession of teddy bears and autism than between autism and MMR.[38] A number of other facts undermined Wakefield's hypothesis: the incidence of autism was already rising before the MMR vaccine was introduced, and autism normally manifests itself at about the age at which children are given the vaccine anyway, which explains the coincidence of timing. Nor was any evidence found to suggest any causal link between the MMR vaccine, bowel disease, and autism.[39] However, these studies, and the failure of Dr Wakefield to find any outside support for his thesis, did nothing to allay public fears.

The scare was the result of a mixture of public ignorance, misjudgement of risk, a general suspicion of science and authority, and the media's love of a sensational story. It is no more realistic to expect tabloid newspapers to eschew sensation in favour of a balanced assessment of risk than to expect predatory carnivores to give up hunting and become vegetarians. But the BBC and the broadsheets acted no more responsibly and were no less cavalier in their treatment of evidence than the popular press. The government's insensitive reaction to public concern did not help. It relied largely on professional reassurances that did little to convince the public, while much greater impact was made by the Prime Minister's refusal to reveal whether his own baby son had received the triple vaccine.

It was an uneven battle. On one side were worried parents with autistic children, who made a highly sympathetic impression when they appeared on television. They included celebrities, an international cricketer, and a popular novelist, essentially conveying the simple message: 'My child is living proof of the harm it (the MMR vaccine) can do.' The mayor of London weighed in with his own contribution to our knowledge of biology: 'It seems to me that a child of 14 months is incredibly vulnerable. Why whack them all (the three vaccines) into a child at the same time?' Dr Wakefield, himself a telegenic figure, came across as someone deeply concerned with the plight of afflicted parents and their autistic children. After a while he ceased to argue with his professional colleagues and concentrated on listening to parents. 'Everything I know about autism,' he said, not entirely accurately, 'I know from listening to parents.' Parents knew because they suffered. He was the picture of concerned humility, the voice of the people, the whistleblower, fighting a battle on behalf of the weak and vulnerable against the mighty, remote, indifferent, unlistening establishment.

On the other side were leading medical experts, quoting statistics that the public, and many journalists for that matter, neither understand nor trust. Epidemiological studies seem particularly difficult for the public and the media to comprehend. Also defending the MMR vaccine were the authorities, the government, and of course the giant pharmaceutical companies, who were automatically accused of existing to make profits out of vaccines. Experts, government, and multinational companies were all natural suspects of a conspiracy and a cover-up. As usual, the history of BSE was brought in to demonstrate that there were good reasons for not trusting experts or official reassurances. No wonder the media were on the side of the parents.

The outcome of the MMR story is that the number of children being vaccinated has dropped below the level judged necessary to provide immunity for society as a whole; in time a measles epidemic may well recur and, if it does, it could cause lasting damage to a substantial number of children and even a number of deaths, all of which would have been avoidable. The Irish experience gives some indication of what could happen. After a measles epidemic in 1999–2000, in Dublin alone there were 111 children in hospital, 12 of them in intensive care, 7 had to be ventilated and three died.[40] Meanwhile, thousands of parents have paid considerable sums to have their children given three separate vaccines that are more safely combined into one. The government is sufficiently confident of the safety of combining several vaccines into one dose that it has added two more vaccines, including one against polio, to the triple dose.

It should be mentioned that one contributory factor to the anti-MMR debate is an underlying suspicion, reflected in the mayor's comments, that vaccines, particularly several vaccines administered together, can somehow overload 'the immune system'. Research suggests that, theoretically, infants can tolerate at least 10,000 vaccines at any one time without their general immunity to infections being threatened.[41] However, the immune system as a concept has acquired a special significance for the general public. Whereas immunologists think in terms of specific immune reactions in which specialized cells are involved in responses to infections, 'the immune system' as such has become an almost mystical entity, constantly threatened by the effects of food additives and pesticides, for example, and by electromagnetic fields and various forms of atmospheric pollution. The concept is strongly promoted by practitioners of alternative medicine, as well as by environmentalists, and has become part of popular pseudo-science. As one psychiatrist has noted: 'The phrase "overload" is frequently used, to portray the idea of the body, or more particularly its immune system, collapsing under the strain of these environmental insults, and hence paving the way for illness'.[42] This concept is part of a general belief that our natural defences are in some way being damaged by human intervention in nature, in which antibiotics and vaccines play an important part.

At the time of writing, the results of the MMR scare are still uncertain, but it is likely that since the level of vaccination against measles has dropped below that required to provide herd immunity, there will be renewed outbreaks of measles with serious consequences for the health of a number of children. It is also likely that, because of the abuse they have incurred, pharmaceutical companies will be less likely to invest in the production of new vaccines and may even abandon the production of existing ones when patents expire.

The removal of tissues and internal organs from dead children

Another example of irrational behaviour is the recent outcry in the UK about the removal of tissues and internal organs from dead babies. In 2001 the public learnt that after autopsy internal organs of many thousands of children had been retained in the pathology departments of a number of hospitals for research purposes, without the knowledge or consent of parents.[43] The reaction was one of public outrage, particularly about a pathologist at the Alder Hey Hospital, Liverpool, who had removed and retained over two thousand organs from hundreds of dead children. The then Secretary of State for Health described it as one of the most shocking events he had ever heard of. All sections of the press echoed the sense of horror expressed by the government. There followed a movement by the parents of the children concerned to recover the missing organs, and in some cases these have since been interred with the rest of the child's body, or buried separately. Invariably, the internal organs were referred to by the press as 'body parts', as if the child's body had been buried in pieces. A law suit was launched by some parents, which was settled by payment of £5 million by way of compensation out of National Health Service funds. Part of the terms of the settlement was that a memorial should be erected to the children. Another group of parents called PITY II ('Parents Interring Their Young Twice') did not claim compensation but wanted a promise from the government 'that organ retention would never happen again'.[44] One lawsuit on behalf of over 2000 relatives is still in train. In an interim ruling in 2003, a judge capped the plaintiffs' litigation costs at £500,000 (compared with their claim that they would amount to £1 million). Costs of £1.45 million had already been incurred. It is a bizarre and sorry tale.

The uproar and reaction over 'the body parts scandal' has had deeply damaging effects on medical research and the recruitment of pathologists, especially paediatric pathologists. The number of post-mortems has been dramatically reduced because families no longer give their consent, fearing that their relatives' organs will be spirited away, while doctors have become shy of asking their permission. Permission to perform a post-mortem examination is given now in only 5 per cent of deaths in a hospital in Sheffield compared with 40 per cent in the past.[45] Three senior brain pathologists told the New Scientist that they used to examine 700 brains a year; in the year after the Alder Hey 'scandal', they received only just over 30. Yet post-mortems are vital to medical research. Without such investigation, relatives often will not know why their children died. Families may be less likely to learn about heritable diseases. Doctors will have less experience to enable them to spot newly emerging diseases or to understand existing diseases better. The value of autopsies can be illustrated by the fact that some types of heart surgery used to be followed by a death rate of one in three, but post-mortem analysis is thought to have reduced the rate to one in thirty-three.[46] Since the Alder Hey episode, three out of five trainee posts in paediatric pathology remain vacant and there is a shortage of pathologists throughout the country. This means that there is a longer wait before biopsy samples from patients are examined. Waiting lists for organs for transplants are also becoming longer. Furthermore, the course of justice has been prejudiced by a shortage of forensic pathologists. These play a vital role in court proceedings, as has been demonstrated in a number of tragic cases when mothers were wrongly accused of murder after multiple cot deaths. (At the time of writing, the courts have ordered a review of a series of cases where mothers in such cases were convicted on the basis of the evidence of one particular expert whose evidence has now been ruled to be unreliable.)

The behaviour of the particular pathologist at Alder Hey was undoubtedly eccentric and could be called disgraceful. But none of the press comment placed the incident in perspective. The retention and storage of organs and histological slides was normal practice. Nothing illegal was done, because the only authority for a post-mortem required by the law in force at the time was that of the person lawfully in possession of the body, namely the hospital manager. Indeed many medical schools were proud of their pathology museums. It was customary for pathologists only to perform an autopsy if a consent form was signed by a relative because they did not wish to be insensitive to the feelings of bereaved families. In practice, parents who were approached considerately were generally ready to give their consent to the use of samples from their dead children's bodies for purposes of research.[47] Not to ask for consent was not a crime, although it was inconsiderate behaviour. However, the subsequent public reaction, spurred on by the press and aggravated by the Secretary of State's comments, implied that the removal of specimens from a dead body is as great a crime as the dismemberment of a living child. Indeed some pathologists even received hate mail that accused them of murdering children.

The legal basis for the action brought by relatives includes a claim for negligence, breach of statutory duty, interference with a body, and infringement of human rights. No one has stopped to ask what possible rational grounds there can be for awarding damages at all. No deaths or physical injury were caused by the retention of tissues. How can hurt feelings of relatives be assuaged by monetary compensation, say by a new motor car or a more luxurious holiday abroad? Yet to dare to question the outcry about the 'body parts scandal' is almost to commit sacrilege. Burial rites are, of course, an old established observance and play an important part in allowing mourning relatives and friends to show respect for the dead and express their grief. However, it seems we have now gone back to the primitive rituals of pre-classical times when bodies had to be prepared for the journey across Acheron into Hades. It is as if no burial can be complete and our human rights are infringed if any part of a body is missing. This is not a case of 'back-to-nature', or 'back-to-the-Middle Ages', but back some 2500 years to pre-classical times.

The dangers of bad medicine

If anyone believes that the flight from reason and indifference to evidence or irrationality in medicine are matters of little importance, the experience of AIDS in South Africa, of MMR and the 'body parts' scandal show how wrong they are. The dangers of alternative medicine can perhaps be mitigated by regulation. But no laws can regulate a desire to return to primitive rites of burial or other forms of irrational behaviour. Perhaps all we can hope for is that certain kinds of medical or scientific claims should be generally recognized to carry a health warning, placing the media and the public on notice that such claims should never be taken at face value. Robert Park, the author of *Voodoo science*, lists seven warning signs which are not, of course, only relevant to medical science:[48]

1. *The discoverer pitches his claim directly to the media.*
   In fact the integrity of science rests on the willingness of scientists to expose new ideas to the scrutiny of their peers. Classic examples of failure to do so are the claims made for cold fusion and the scare first raised on British television that genetically modified potatoes could be a cause of cancer.

2. *The discoverer says that a powerful establishment is trying to suppress his or her work.*
   The evidence produced to challenge Dr Wakefield's claims about MMR and autism was denounced by the campaign against MMR as a conspiracy against him by the medical establishment, the government, and the pharmaceutical companies. This distracted attention from the flaws in his own research.

3. *The scientific effect involved is at the very limit of detection.*
   Compounds used in homeopathy are indeed totally undetectable after their massive dilution in water.

4. *Evidence for a discovery is anecdotal.*
   Anecdotes keep superstitious beliefs alive. The need for double-blind tests, which tell us what works and what does not, is ignored. As Park observes, ' "Data" is not the plural of "anecdote".'

5. *The discoverer says a belief is credible because it has endured for centuries.*
   This applies to all 'traditional' treatments; ancient Chinese lore is particularly favoured. The history of medicine hardly proves that what is old must be good.

6. *The discoverer has worked in isolation.*
   The lone genius is more often found in films than real life.

7. *The discoverer must propose new laws of nature to explain an observation.*
   Again, this is common in alternative medicine; for example, the basic principles of homeopathy cannot be explained by any known laws. Other practices depend upon the existence

of different systems of physiology or anatomy unrecognized by science.

However, there is a further objection to irrationality in medicine, which forms part of the theme of this book. It is dangerous for society, as well as medicine, if people are encouraged to turn their backs on reason. The seller of snake oil, of *Elisir d'Amore*, or other quack remedies is not a harmless crank. He is a con-man who sets out to deceive. Practitioners of homeopathy are not sellers of snake oil, because they do not seek to deceive their patients deliberately and are no doubt convinced, like doctors who administered purges in the past, that they are healing the sick. Indeed, sometimes they are right, because their potions may help or provide comfort through the placebo effect. However, those who preach and practise homeopathy encourage people to believe in nonsense. The law of infinitesimals is patent nonsense. It is as irrational as dependence on miracle cures and the practice of Ayurveda. As Christopher Wanjek points out in his book *Bad medicine*: 'When you place your trust in a proponent of Ayurveda, you are also placing your trust in someone who likely claims to be able to levitate, read minds, foretell the future, reduce crime and end war through meditation, or heal with chanting, cow dung, and spit'.[49] If you abandon any concern for evidence or pretence at reason, you open the door wide to more dangerous charlatans, the peddlers of racial hatred, or those other devotees of the irrational, the religious fundamentalists who seek a return to the days when religious dogmatism ruled and freedom of thought was suppressed. To ignore evidence which contradicts or casts doubts on your claims is to show indifference to the truth. Truth matters and disregard for the truth weakens the moral quality of civilization. Belief in irrational medical treatment may not be the most dangerous form of irrationality, but it is still a step in the wrong direction along a fearsome road.

3
The Myth of Organic Farming

> Organic farming is sustainable. It sustains poverty and malnutrition.
>
> C. S. Prakash, distinguished plant biologist

ACCORDING to the Oxford English Dictionary, 'organic' describes compound substances that naturally exist as constituents of animals and plants. All food is organic. It has to be, because all animals, including human beings, are themselves organic and have evolved to digest organic matter. Non-organic, or inorganic, farming is therefore an oxymoron and the phrase 'organic farming' is a meaningless phrase, essentially tautologous. However, the term 'organic' has been appropriated by the followers of a particular movement and given a specialized meaning. Farming and its products only qualify as officially 'organic' if they comply with certain rules and principles. (The rules prohibit the use of most artificial fertilizers and pesticides and animals are to be kept in ways that minimize the need for medicines and other chemical treatments.) Originally based on a particular philosophy of life concerned with man's place in nature that first emerged in Germany in the early twentieth century, these rules are now laid down by a number of certifying bodies. In Britain the main ones are the Soil Association, the voice of organic farming in Britain, and the United Kingdom Register of Organic Food Standards (UKROFS). Outside the United Kingdom, the main controlling body is the International Federation of Organic Agricultural Movements.

Nowadays 'organic farming' commands such wide public support that to question its merits is to question the virtues of motherhood. Nearly every famous chef, cookery expert, and item about food in lifestyle magazines or on television, takes it for granted that organic food tastes better and is more nutritious and better for our health. Nearly every environmentalist is convinced that organic farming is better for the environment. We are constantly told that it preserves the fertility of the soil, prevents pollution of the water supply by nitrates, and that it will reverse the decline in biodiversity, especially in populations of farmland birds. The British Government subsidizes farmers to convert to organic farming, and in 2002 an official policy commission on Farming and Food[1] recommended that even more money should be spent to ensure that organic farming plays a larger role in agriculture. In Germany, the Minister for Agriculture at the time of writing, Renate Künast, is a member of the Green Party and has declared her objective to be the maximization of organic farming in the European Union. (However, fellow Greens in Britain and elsewhere must have been surprised when she announced, in January 2004, that the German Government would licence the commercial planting of GM crops and that she saw no health risk to consumers). Other EU agricultural ministers seem only too ready to follow her lead where organic farming is concerned and they envisage that in due course over 20 per cent of European agriculture will be organic. Throughout Europe organic farming is expanding annually at rates of up to 20 per cent. However, the figures for growth give a somewhat misleading picture. Not only do they start from a low base, but much of the expansion relates to grassland for feeding sheep and cattle, which requires no special treatment. In the UK the proportion of vegetables grown organically is only 0.5 per cent.

To the ordinary public, the label 'organic' has a reassuring ring, particularly when contrasted, as it constantly is, with 'synthetic'. Eating 'organic' food is like drinking 'real' ale, not *ersatz*, imported, imitation stuff. It sounds safe because it is guaranteed to be GM-free and is assumed to be untainted by nasty, possibly carcinogenic pesticides. Supermarkets promote it, which they would not do unless there were a popular demand for it; it is also clearly to their advantage that the public are prepared to pay premium prices for it. More and more farmers look to a future in organic farming because its higher prices offer the prospect of higher profits, one bright spot in the otherwise bleak landscape of the agricultural industry in Europe. In fact, domestic supply in Britain cannot keep up with demand, so that over 70 per cent of organic produce has to be imported. (As one of the advertised attractions of organic food is its freshness, clearly most organic food on supermarket shelves does not qualify. Indeed, if one takes into account the air miles flown to bring organic food to European markets, most organic food in the shops cannot be regarded as environmentally friendly.)

Perversely, evidence to justify public enthusiasm has proved elusive. The Food Standards Agency (FSA) in Britain, set up to examine evidence about the safety of food and to protect the interests of consumers, has persistently refused to uphold claims for the superiority of organic food, much to the chagrin of the Soil Association. In January 2004 the FSA stated: 'On the basis of current evidence, the Agency's assessment is that organic food is not significantly different in terms of food safety and nutrition from food produced conventionally'.[2] When a complaint was made to the Advertising Standards Authority that recruiting leaflets published by the Soil Association made misleading statements, claiming that organic food tastes better, is healthier, and is better for the environment, the Authority found no convincing evidence to support the claims and the leaflets had to be withdrawn.[3]

It is not surprising that these two independent bodies should find no evidence to support the claims, because in fact public faith in organic food is based on myth. The organic movement has murky origins; its basic principle is founded on a scientific howler; it is governed by rules that have no rhyme or reason; it is steeped in mysticism and pseudo-science; and, whenever it seeks to make a scientific case for itself, the science is shown to be flawed. If organic farming were to be much more widely practised, as its supporters advocate, it would have damaging consequences for farming as a whole, for the world food supply, and for the environment.

Myths and mysticism

The Soil Association was founded in 1945 to promote non-intensive farming methods that preserved the structure and fertility of the soil. Its first President was Lady Eve Balfour, who believed that vital principles were found in manure and that plants grown in manure generated healthier food than that produced by the application of minerals. But the original inspiration for organic farming came from the early twentieth century mystical philosopher Rudolf Steiner, a follower of the German *Naturphilosophen* (e.g. Fichte, Schelling, and others) of the nineteenth century. This was a group whose obscurity of language was exceeded only by the obscurity of its ideas. Indeed, one of its most celebrated philosophers, Friedrich Schelling, averred that 'it is a poor objection to a philosopher that he is unintelligible'.[4] In his lectures on agriculture in the 1920s, Steiner stressed the virtues of manure as a soil fertilizer. He believed that cosmic forces entered animals like cows or stags through their horns, and he developed a concept of feeding the soil through a process of 'biodynamic cultivation',[5] which involved planting according to the phases of the moon and nourishing the soil with cow horns stuffed with entrails. He also taught that chemical fertilizers damaged the human nervous system and the brain.[6]

The mystical origins of the organic movement would be irrelevant if the Soil Association, the main promoter, controller, and defender of organic farming in Britain, did not regularly dismiss scientific criticism by stressing the need to look beyond science to the spiritual or mystical dimensions that farming should take into account. The Director of the Soil Association, Patrick Holden, has dismissed the idea that the achievements of organic farming could or should be scientifically tested, because organic farming is 'holistic, integrated and [represents] joined-up thinking'. The trouble with asking for scientifically based measurements is that the organic, holistic approach is not 'reductionist'. He has deplored the 'obsession with reductionist science: ... holistic science strays into territory where the current tools of understanding that are available to the scientific community are not sufficiently well developed to measure what is going on'.[7]

Holden's statement that current science is not sufficiently developed to evaluate organic farming echoes almost exactly comments made by the editor of *Alternative Therapies* (see Chapter 2, p. 43) to the effect that the intrinsic qualities of alternative medicine cannot be measured by contemporary scientific methods. Rejecting the methods of science as 'reductionist' makes assessment of the effectiveness of organic farming impossible, because only by changing one factor or variable at a time can cause be related to effect. But the organic farming lobby, like supporters of alternative medicine, do not believe in the scientific method. Both practices have virtues, it seems, that can only be detected by intuition; they are both revealed as based on a belief in magic or mysticism, not reason.

A lack of concern for scientific evidence, indeed for simple facts, is also evident in the basic credo of the contemporary organic movement, which is the belief that synthetic chemicals are bad and natural chemicals are good. This belief inspires the rules of the movement and pervades the writings of its devotees. It is an extraordinary belief. First of all, it ignores the fact that a molecule is a molecule; the product is the same, whether it is made by a man-made synthetic process or by a natural one. Secondly, it denies elementary chemical truths: that many synthetic chemicals are beneficial. Conversely, many natural chemicals can be poisonous. Anti-bacterial drugs like sulphonamides or isoniazid, which kills the tubercle bacillus, are synthetic. So is the painkiller paracetamol. Poisonous chemicals found in nature include ricin, aflatoxin, and botulinum toxin. In every case, whether the chemical is beneficial, harmless, or harmful will, as the Swiss Renaissance physician Paracelsus observed centuries ago, depend on the dose. Too much of anything, including water, will kill you; very small doses of arsenic do no harm, and indeed there is evidence that they can actually do good. The belief in the goodness of what is natural and the sinfulness of what is man-made is part of the 'back-to-nature' philosophy that regards science, and its attempts to control or improve on nature, as one of the baneful influences on humankind. It overlooks the fact that cholera, plague, starvation, and any number of other scourges of humankind were afflictions of nature that synthetic medicines and technical advances have enabled us to control.

It is therefore clear that the leaders of the organic movement on the whole do not care about scientific comparisons and prefer intuition and mysticism, and, not surprisingly, are happy to ignore elementary chemistry to base their doctrine on a false distinction between natural and synthetic chemicals. But does their devotion to mysticism and indifference to science necessarily discredit the whole organic movement? Since farming only qualifies as organic if it complies with rules made by the Soil Association or by UKROFS, perhaps the most important questions are whether these rules make sense and whether, in practice, farming in accordance with them has the merits claimed.

Rules with no rhyme or reason

Unfortunately, the rules themselves are inconsistent, arbitrary, and reveal no coherent set of principles. The use of some pesticides is allowed, for example spraying with *Bacillus thuringiensis (Bt)*. This is the same *Bt* bacterium whose insect-resistant genes have been transferred to maize, soya, cotton, and other genetically modified plants, yet the Soil Association is one of the principal Green lobbyists campaigning against its use in GM crops. The official position of the organic movement, confirmed by its rules, is that the presence of a particular *Bt* protein within a plant as the result of genetic modification is dangerous, but the organic farmer can spray the plant with *Bt* spores containing that same protein. In both cases, the bacterial *Bt* protein protects the plant from its insect pests. Nothing could more clearly illustrate the topsy-turviness of the Soil Association's make-believe world.

However, when crops are genetically modified to incorporate a *Bt* gene, particular pests are specifically targeted by the insertion into the plant of one or two genes that code for the toxic protein that affects those pests and no others, so that the minimum amount of harm is done to other non-target insects or to natural predators of that pest. For example, the gene for the *Bt* protein that kills one species of caterpillar is used in plants that are attacked by that caterpillar and the gene for another toxic protein that kills a particular beetle larva is used in plants for which those larvae are the main pest. By contrast, when the organic farmer sprays *Bt* spores onto his crops, the spray contains a mixture of toxins, since the *Bt* bacterium produces some 130 different toxins, each of which is active against a particular kind of insect. Such sprays are not specific in their effect. They are more likely to affect non-target insects (i.e. beneficial insects) than the toxic protein expressed by a *Bt* gene in a GM plant. The organic farmer also has to spray repeatedly, which is expensive; transgenic *Bt* plants do not have to be sprayed. Thus, the Soil Association rules in this case explicitly discourage the better environmental practice.

Another arbitrary rule permits the use of the inorganic compound copper sulphate as a fungicide. Although the use of copper compounds in agriculture was due to be prohibited across the European Union from March 2002, limited use has been permitted until 2006 at the express request of the organic movement. Why do the organic rules allow the use of copper fungicides on potatoes, when they prohibit the use of better, well-researched, and safer fungicides? Copper-based fungicides are less effective against late blight, and are more toxic to insects, than any of the more modern classes of fungicides. They are also more persistent in the environment and more damaging to the soil. The only reason for a plea for their continued use seems to be that they are the oldest in regular use and are venerated because they are traditional.[8]

Even if its rules are illogical, contradictory, and arbitrary and even if the central philosophy of the movement itself is based on a fundamental scientific error, it is still possible that, by accident as it were, organic farming actually works and that its effects are beneficial. The public, and, one suspects, most organic farmers, do not care about the philosophy behind the rules, and few will have heard of Rudolf Steiner. But people clearly see practical merits in organic food, since they buy it even though it costs more. A survey in 1997 showed that 83 per cent of consumers bought organic food to avoid pesticides; 75 per cent on the grounds that it is kinder to the environment; 70 per cent were concerned about the intensive rearing of animals; 68 per cent bought it because of the taste; and 36 per cent expressed worries about BSE. Since that survey, another commonly expressed concern is about GM food: consumers buy organic food because it is GM-free.[9] Surely people 'know their onions' and if they like it and are prepared to pay a higher price for it, it must have some merit?

Each of these reasons will be considered separately. But the fact that people buy it is no more proof of its merits than the fact that most people's belief in it proves the merits of astrology, or homeopathy, or that there is a link between MMR and autism. The philosophical reasons for supporting organic farming are part of the 'back-to-nature' syndrome. Like the practice of alternative medicine, they are based on the belief that 'nature knows best' and that what is natural must be good. It is a belief that betrays a certain nostalgia for a mythical golden age of small-scale and simple farming and pure and wholesome farm produce, before modern technology interfered with nature and spoilt the Arcadian countryside. Such a paradise never existed. In the days before intensive farming, when farmers did not use pesticides or artificial fertilizers, food supplies were constantly endangered through climatic and environmental fluctuations and crops were frequently lost to pests and diseases. Agriculture was associated with grinding poverty, intensive labour, and low yield. The poor quality of much food, together with infectious diseases, contributed to a much shorter life span of the general population. Malnourishment was rife. In Britain, for example, 60 per cent of potential recruits for service in the Boer War, in 1900, were rejected by the army because of their low stature and weight, as the result of an inadequate diet.

In the last fifty years, since synthetic chemicals came to be widely used, our life expectancy has increased by seven years or more. Healthier and safer food, together with better health provision, has improved our physical well-being and increased longevity, and modern agriculture deserves much of the credit.[10]

The virtues of 'natural farming' and the 'back-to-nature' cult appeal strongly to the media, who treat the Soil Association as an authority deserving at least as much respect as the Royal Society. After all, the organic people are the good guys trying to give us wholesome food and save the countryside. When, therefore, the Soil Association produces research it has commissioned to justify its claims, no interviewer ever asks if there is any independent verification. But if we want to know how organic food compares with other food, we need objective comparisons that compare like with like. It is only too easy to parade specious comparisons that are superficially persuasive but totally misleading. Farms vary enormously in different parts of the UK, let alone in different parts of the world. Wind and rainfall vary, so do the soil, the hedgerow structure, the weeds, and pests, and all of them affect the efficiency and environmental impacts of a farm. How the produce from one farm compares with another also depends on the quality of management, which is probably the most important factor that affects the impact of a farm on the environment. If someone sets out to farm in an environmentally friendly way, it is likely that he or she will succeed. Indeed it is because many people take up organic farming for environmental reasons that many organic farms have a good record for promoting birdlife and biodiversity. But the same results can be obtained by other farming systems, if they too are managed with the same dedication. According to the Rothamsted Research Institute, 'where one tries to match the farm type, the butterfly and bird numbers can be as good on a con-ventional farm as on an organic farm'.[11] Proper comparisons should therefore be between organic and conventional plots farmed by the same farmer. Fortunately, several such comparisons have been made and, moreover, they have compared performance over a sufficiently long period of time to eliminate accidental factors.

Does organic food taste better?

As polls show, most people believe it does. In blind tests, however, there is a common confusion between organic produce and fresh-ness, and the public has not been able to distinguish organic from conventional food.[12] Such scientific tests produce a result at such variance with so many people's declared experience, including that of many food experts, that it seems to require some explan-ation. One reason may well be a common confusion with freshness.

Organic food often tastes better because most home-grown organic products are fresher, for the simple reason that they have a short shelf-life. In the case of chickens, there is some confusion between organic and free-range: many people assume that free-range chickens must be organically reared. Again, local variables can produce different results, because of differences in the soil, weather, and management practices. For example, in a comparative study of different farming systems at Boarded Barns at Ongar in Essex, a panel carrying out blind tests found that organically pro-duced bread had a mustier taste and did not taste as fresh as bread from conventionally produced grain or that produced by inte-grated farm management.[13] The fact remains that the Advertising Standards Authority, with no vested interest in its conclusions, found the claim that organic food tastes better was not supported by evidence and academic studies came to the same conclusion.

Is organic food healthier?

This is one of the most important questions since the main reason people give for buying organic food is to avoid pesticide residues. The Soil Association plays on these concerns, as do a number of other campaigning organizations that have helped to create a food-scare industry. For example, in November 1998 the Con-sumers' Association magazine *Which?*, under the heading 'Pesticide Concerns', carried a story that test results from animal studies linked high doses of pesticides with cancers, hormone disturbances, and birth defects. It did not mention that high doses of anything cause harm, or that official reports on the concentrations of pesticide residues in food found that the amounts present were so low as not to constitute a hazard to health.

A typical example of the case made against the use of pesticides was a detailed indictment published by a leading figure in the Soil Association in *The Guardian*.[14] She complained that pesticide residues 'have become a routine ingredient in our diet'. In the year 2000, '67 per cent of the grapes, 72 per cent of the apples and 71 per cent of the pears we ate contained residues. ... As a working rule of thumb, at least 40 per cent of all the fresh fruit and vegetables we eat contains residues, often multiple residues, of several pesticides and, not infrequently, illegal ones'. She acknowledged that fewer pesticide residues were found in the UK than in other countries, but suggested this was because our system of monitoring was less rigorous. She also conceded that an Advisory Committee regulates the pesticides that may be used by growers, and that its chairman is independent, but she noted that several of its members have done work or acted as consultants for chemical and biotechnology companies and inferred that the committee therefore has a vested interest in approving pesticides. Her conclusion was that the Advisory Committee is in the pocket of the companies, who, it seems, are quite happy to poison us for the sake of profits.

Some of the residues in our food, her article revealed, are of chemicals like organophosphates, 'infamous for their devastating effects on the central nervous system.' We should not only be concerned with the effect of residues in pears, apples, and grapes already mentioned, but also strawberries, peppers, and chocolates, spinach, celery, carrots, oranges, potatoes, oily fish, and wholemeal and multigrain bread. The Government's rationale for its approval system, that huge safety margins are built in, was dismissed out of hand, as was the idea that the public is not at risk if there is no evidence of harm. As usual the BSE experience was cited to show how mistaken this approach has proved. The Government is also at fault, she maintained, for not giving the control bodies a remit to encourage organic farming. Finally, the conclusion was reached that consumers are left with two principal choices: 'You can switch to organic ... (her association with the Soil Association is not mentioned). Or you could just accept that every third mouthful of food you eat contains poison. Are you up for that?' As this kind of alarmism is not uncommon, it is not surprising that 86 per cent of consumers wish to avoid all pesticide residues.

Now it might, at first sight, seem sensible to ensure if possible that there is no residue at all of anything poisonous in food. But the writer, like all pro-organic anti-pesticide campaigners, forgot the message of Paracelsus: it all depends on the dose. At no time does she mention the concentration of any of the residues found. Detection itself is not enough to justify expressions of horror. If it were, warning us that one mouthful in three contains poison is not being nearly alarmist enough. In fact *every* mouthful of food contains some poison, as does every sip of water. 'Carcinogenic' substances are routinely consumed by all of us in the form of natural chemicals made by plants to repel predators, but amounts are so small they do not harm us. Potentially harmful chemicals including arsenic are found in many foods and in drinking water, but the quantities are, usually, too small to cause harm. There are some dioxins in every breath of air we take, but again in such small amounts as to be insignificant. In fact they may actually do good (see p. 72).

It is worth quoting a review by Sir John Krebs, chairman of the Food Standards Agency, published in *Nature* (a journal in which inaccurate or unfounded statements are seldom left uncorrected): 'A single cup of coffee contains natural carcinogens equal at least to a year's worth of carcinogenic synthetic residues in the diet'. He points out the disparity between public fears about food and the facts:

dietary contributions to cardiovascular disease and to cancer ... probably account for more than 100,000 deaths per year in Britain. Food poisoning probably accounts for between 50 and 300 ... pesticides in food, as well as GM food, are not responsible for any deaths.[15]

The distinguished microbiologist Bruce Ames states that Americans eat 1500 milligrams of natural pesticides a day, an amount about 10,000 times greater than their daily consumption of synthetic pesticide residues.[16]

One reason why the public is acutely conscious of pesticide residues in food is that we have become much better at measuring the very small amounts present. As most people cannot distinguish between micrograms and picograms, more sensitive tests, which should provide reassurance, paradoxically frighten people instead.

Of course public concern about pesticides is not, I believe, only due to anti-pesticide propaganda from the organic movement. It is part of a phobia about carcinogens for which Rachel Carson also bears responsibility, through her claim that organochlorines such as DDT caused cancer. Today there is a widespread belief that there is an epidemic of cancer caused by various forms of environmental pollution, including pesticides. In fact, most forms of cancer are associated with smoking, obesity, and sunshine and are otherwise connected with the fact that we live longer. Overall, cancer rates are in decline, particularly when lung cancer induced by smoking is removed from the detailed age-related statistics.[17] It is significant that cancer rates among farmers are about half the average, although farmers are more exposed to pesticides than the rest of us. It is also interesting that the incidence of cancer of the stomach, which is likely to be related to diet, has declined by 60 per cent in the last fifty years, a period during which the use of pesticides in agriculture has increased.[18] Fear of pesticide residues in food is one more example of a health scare without foundation.

Low-dose beneficial effects or the 'hormesis' effect

Ironically, there is persuasive evidence that low concentrations of many toxic chemicals may actually have a beneficial effect. The phenomenon of hormesis, or low-dose beneficial effects, is widely observed and accepted.[19] It seems that the hazards of low-level exposure to pesticides may have been overestimated and scientific and regulatory approaches to pesticide management are being reconsidered by toxicologists.

Examples are, of course, familiar. A small dose of aspirin mitigates a headache and can help prevent heart attacks, but a larger dose can kill. Fluoride in small doses strengthens teeth and bones, but it is a poison. Sunshine is good for us if we protect ourselves against overexposure, but causes melanomas and other skin cancers if we do not. A little bit of dirt helps stimulate your immune system. Most encouraging of all, moderate consumption of wine protects against cancer and cardiovascular disease, although overindulgence can be fatal. It is not generally realized that this dose-related effect called 'hormesis' is also known to apply to many supposedly toxic chemicals, including arsenic, dioxins, some pesticides and fungicides—and even diluted factory effluent and radiation.[20] In fact, a little bit of poison or pollution can do you good, and serves to reduce the incidence of cancer. Over 30 separate investigations of about 500,000 people have shown that farmers, millers, pesticide-users, and foresters, occupationally exposed to much higher levels of pesticide than the general public, have much lower rates of cancer overall.

By demanding total elimination of all pesticide residues from our fruit and vegetables, the organic movement promotes an unreasonable fear of chemicals and scares us about non-existent dangers. The public is not made aware of their beneficial effect on our general health.[21]

Is organic farming better for the environment?

Another reason given for buying organic food, to some its main attraction, is that organic farming is friendlier to the environment. Many people buy organic for the same reason that they recycle paper and glass: they feel that they are being responsible citizens and are doing their bit to preserve birds and butterflies. Organic farms do show environmental benefits, in that more birds and butterflies as well as other insects inhabit them than most conventionally farmed land. Indeed, the idealism that makes many people take up organic farming should not be discounted. They want to preserve and encourage biodiversity and believe that organic farming is the answer. One of the virtues of their rules is that UKROFS and the Soil Association specifically require organic farmers to aim for environmental benefits, to maintain soil fertility, rotate crops, avoid pollution, and show concern for animal welfare. Organic farmers set out to manage their farms to achieve good environmental effects and it is not surprising that they do so.[22]

Because I argue that organic farming has no scientific basis and has many disadvantages, I want to make it clear that I admire the achievements of many small organic farmers in improving the environment. I share their aims and indeed many of their dislikes. Factory farming, for example, of chickens and livestock is a deeply repulsive practice and, in the balance we have to strike between the economic interests of human beings and respect for nature and its creatures, I regard the low prices of poultry and meat, which we owe to factory farming, as too high a price to pay. I agree with the aim of the organic movement to reverse the damage some of the practices of intensive farming has caused to biodiversity.

However, as the evidence demonstrates, this is a matter of management, not of the system, and it can be achieved by other means than organic farming. The effect of different farming systems on the environment was tested at Boarded Barns in Essex in a meticulously conducted comparison of organic farming, conventional farming, and integrated farm management (IFM)—a system that specifies exacting standards of landscape, hedgerow maintenance, large field margins, and insists on high standards of animal welfare. Indeed, IFM incorporates all the attractive features of organic farming without its ideological absurdities. The study was sponsored by Aventis, but the work was done by a number of independent universities, institutes, and environmental organizations, including the British Trust for Ornithology, the Essex Farming and Wildlife Advisory Group, and the Essex Birdwatching Society.[23] The effect of the different systems was compared over a ten-year period, an important feature, since it takes many years to assess the effects of changes in agricultural practice.

The report listed as its most important finding that the particular farming system used had less direct impact on key areas of biodiversity than was earlier supposed. Overall, the best results came from IFM, many of whose techniques have now become common practice for conventional farmers. By most environmental tests—soil quality, effect on bird life, numbers of mammals and insects—it scored at least as well as organic farming, and overall it was the best in terms of biodiversity. It also required less fuel and was more efficient in its use of labour than organic farming. The latter was superior in only one respect: the high premium prices for organic food made it more profitable. One of the important findings was that 80–85 per cent of animal life in any farm exists in the field margins and hedgerows and that the effects of pesticide application on the cropped area is of little significance. Thus any system that maintains margins and hedgerows is likely to be as good for biodiversity as any organic field.[24]

What people care about most is the effect of farming on birds. (The Royal Society for the Protection of Birds is one of Britain's richest charities). Yet evidence for the effect on birds of different farming systems is difficult to establish. The main difficulty is that large fluctuations in bird populations have occurred in the past and we do not know why. For example, the recent decline in song birds may well be partly due to cats, whose numbers have increased by 50 per cent in the last 20 years to some 8 million. Domestic cats, it is estimated, are to blame for the deaths of some 300 million young birds and small mammals every year. Some bird populations go up (sparrowhawks), while others go down (tree sparrows). Any decline in a bird population is automatically blamed on intensive farming, while no one has yet suggested that it is responsible for any increase.

A number of studies have also been done on the effect of different systems on soil fertility, soil structure, and on nitrate pollution of waterways, but these too are broadly inconclusive. One verdict of a comprehensive review of the literature is 'that little or no benefits follow from current organic procedures. . . . the supposed destruction and erosion of the soil in Britain [which led Lady Balfour to found the Soil Association] no longer occurs and the case for supporting organic agriculture on this basis is not justified'.[25]

Among the most important causes of damage to the environment, rarely stressed and completely ignored by the organic movement, are the tractor and the plough. On organic farms, weeds are controlled by frequent mechanical weeding. But the tractor and the plough damage worms and insects in the soil, cause soil erosion, release more carbon dioxide into the atmosphere, disturb nesting birds, use more fossil fuels, and are in every way less beneficial to the land than the no-tillage, or low-tillage farming made possible by genetically modified, herbicide-tolerant crops. These, it has been estimated, reduce greenhouse gases by over 80 per cent per hectare (see Chapter 4, p. 104 below). In winter the number of birds on no-tillage fields exceed by many orders of magnitude the number of birds found on organic fields.[26]

Efficiency and the future of farming

Possibly the most telling indictment of organic farming is its inefficiency—its high cost and its wasteful use of land. The facts cannot be seriously disputed. One study purporting to show that organic and conventional corn yields were identical omitted to mention that it required twice as much land to achieve the same yield. In an occasional year yields from organic farms can be equivalent, but since the organic process depends upon a ley period in which clover and grass or alfalfa are grown to allow nitrogen fixation and provide the soil with nitrogen to be ploughed in, total yields have to be compared over a continuous number of years. An experiment which made a valid comparison of yields from organic and conventional produce at the same farm reported that the yields from organic wheat, beans, and peas were only 60–70 per cent, and of oats 85 per cent, of conventional yields.[27] The Boarded Barns study routinely reported that organic wheat yields using animal manure were about 50 per cent of those of conventional wheat. The evidence is overwhelming: yields of most crops from organic farms are about 20–50 per cent lower than from conventional farming. That is why organic food costs more.

Does it matter? The argument frequently advanced by the organic lobby is that we have become obsessed with efficiency. The consumer is sovereign, and if the consumer likes organic food, does it matter if it is less efficiently produced and costs more, since many people are prepared to pay premium prices? If organic farmers can make a good profit and build an enclave of prosperity in a landscape of depression, surely organic farming should be encouraged. If consumers want it, that is justification enough for organic farming.

Efficiency does matter. It affects the health of low-income families. Even in a prosperous society like Britain we should not ignore the importance of cheaper ways of producing food, provided they are not based on intolerable breeding conditions for animals. Prosperous (and vocal) middle-class consumers may not care about price, but the poorer you are, the more the price of food matters. Pesticides keep down the cost of fruit and vegetables and if the organic lobby prevails they will become more expensive. People in the lower-income groups will buy less fruit and fewer vegetables; this is all the more important since they are now exhorted to eat more of them to help control obesity. Moreover, the more pervasive the propaganda that more expensive organic food is 'safer and healthier', the greater the pressures on poorer families to buy food they can ill afford. Their diet will suffer and they will lose the protection against cancer that a healthy diet provides. More of them will die younger compared with the rich. Our model should not be Marie Antoinette making dietary recommendations to hungry Parisians.

Even from the farmers' point of view, it is doubtful whether a system that depends on premium prices paid for food of no superior quality can provide a sound long-term basis for a viable agricultural industry. Today's premium depends on the organic market being a niche market. As the number of organic farmers increases, encouraged by government subsidy, the premium will fade away. In 2001, 18 new organic dairy farms in England came into operation and their produce overwhelmed the small market for organic milk, forcing the price down from 30p to 24p a litre, only a penny more than the price of milk from conventional farms. The new farms produced smaller yields at higher cost and inevitably some organic milk farmers went out of business. The same fate might befall organic farmers growing other crops.

There is also an ethical issue. At the moment, supermarkets benefit from high prices for organic produce. There is an element of deception when companies boost profits by promoting the sales of more expensive products that do not reflect better value. Supermarkets claim they are providing what customers want. However, far from educating their customers to get value for money, they encourage them to buy organic food. Imagine the outrage if multinational agri-business exploited consumers in the same way.

The environment also suffers if farming is inefficient. Organic farming wastes good farmland. Since Europe produces an excess of food as a result of efficient farming, farmers can be encouraged to set aside half their land for environmental purposes, for woodland or fast growing willow plantations which can be coppiced frequently and the wood used as fuel. Such plantations, with their undercover of weeds, bird-nesting sites, and mammal and insect refuges, are more effective at promoting biodiversity than any organic farm, use less fossil fuel, and produce much less carbon dioxide. They are already a common feature in many European countries.[28]

However, all these considerations are minor compared with the needs of the world as a whole. The poorest farmers in Africa and Asia are already organic farmers: they do not use pesticides or artificial fertilizers because they cannot afford them. The Green Revolution passed them by, which was one of its failures. The organic movement seeks to go back to the days before the Green Revolution. It cannot help eliminate the pests and diseases that destroy nearly half the crops in Africa, or the development of drought-resistant crops that can grow on arid or semi-arid land. It cannot even match the yields which conventional farming already achieves today. What is more, in many parts of the world the only way in which inefficient organic farmers can feed a growing population is by cutting down more tropical forest: for example, Mexican farmers currently 'slash and burn' three million acres of virgin tropical forest a year.[29] Organic farming may satisfy the whim of the rich European or American consumer; its extension to the developing world would be a disaster. As the Indian bio-technologist, C. S. Prakash,[30] has correctly observed: 'The only thing sustainable about organic farming in the developing world is that it sustains poverty and malnutrition'.

Scientists, who know that there is no intellectual case for organic farming and who are fully aware that its principles are based on myths and untruths, frequently say they have nothing against it: good luck to the farmers who make profits from it and to consumers who are happy to buy it. I believe this position is morally untenable. Truth matters, and if an important industrial activity is based on nonsense we should say so. We should not encourage superstition but expose it. When medicine is based on voodoo science, the danger is not only to the health of patients who may be misled, but to the way we approach the problems of life. Organic farming is based on pseudo-science and it is import-ant that this should be publicly recognized. One of the main pur-poses of education is to teach children to think straight and to distinguish the true from the false. Woolly thinking about food and farming is as much a manifestation of unreason as belief in homeopathy.

Nor should we be indifferent to a movement which makes it less likely that poorer families will improve their diet and more likely that they will suffer ill health as a consequence. It is an indefens-ible part of government policy, influenced by the power of the multi-million pound organic farming lobby, to subsidize this harmful nonsense. Above all, protestations that we care about world poverty ring false when prosperous nations protect their own farmers with subsidies and penalize subsistence farmers in the developing world. To promote organic farming and exacerbate the shortage of productive land compounds hypocrisy.

4

The Case for GM Crops

> And he gave it as his opinion, that whoever could make two
> ears of corn or two blades of grass to grow upon a spot of
> ground where only one grew before, would deserve better of
> mankind, and do more essential service to his country than
> the whole race of politicians put together.
>
> Jonathan Swift

THE popularity of alternative medicine and organic food reflects the growing strength of the Back-to-Nature movement, but in neither case has it yet carried the day. After all, most people who are seriously ill still go to a qualified doctor; organic food is not our main source of food and organic farmers produce only just over one per cent of all agricultural crops in the UK.

However, opposition to genetically modified crops by Green lobbies has been so successful, at least in Europe, that the future of agricultural biotechnology here is bleak. At the time of writing, few GM crops have yet been licensed to be grown commercially and virtually no such food is sold in European shops.[1] While the European Commission is eager to encourage biotechnology, including GM crops which it has declared to be safe, EU member states, except Spain, are generally reluctant to license GM crops. The companies that promote them have withdrawn their research activities from Europe to concentrate them in the United States where the political climate is less hostile, with the likely result that when Europe finally accepts GM products, as is inevitable, they will be of American, not European, origin and agricultural biotechnology will not have a European base. Meanwhile Europe's hostility to GM food in turn inhibits the production of GM crops in developing countries that might hope to export their produce to

European markets. It was claimed by opponents to the technology that the strength of public opposition in Britain was revealed by the response to an exercise in public consultation conducted by the British Government in 2003, when 94 per cent of those who were consulted said they would not eat GM food. However, a more thorough survey found that 36 per cent were clearly opposed to GM food, 13 per cent in favour, and 39 per cent uncommitted. Some 85 per cent felt they did not know enough about the potential long-term effects of GM food on our health.[2] The stridency of the opponents gives the impression that opposition is stronger than it really is, but its effect has still been devastating on the future of GM crops in Europe. Supermarkets, for example, think it necessary to assure their customers that their goods are GM-free.

The force of opposition to the technology has also been strong enough to force plant biologists, who overwhelmingly support it, onto the defensive. Academics writing about genetic modification of plants tend to stress their neutrality, as if to reassure us that they know arguments for and against are evenly balanced. A good example of this is an excellent book, ideal as a textbook for schools or as an introduction for the uninformed, *Pandora's picnic basket* by Alan McHughen, a Canadian research scientist.[3] The author announces from the start that he will chart a middle course between the exaggerated claims of the pro-GM and anti-GM camps and between potential risks and benefits. In fact his book demonstrates that there is no even balance between the two extremes. His only example of somewhat exaggerated claims in favour of genetic modification is an ill-fated advertising campaign by the multinational company Monsanto in Europe in 1999—which became notorious and will be remembered in perpetuity—whereas examples of false scares can be identified almost daily.[4] He is equally hard-pushed to name demonstrated (as opposed to theoretical) risks, whereas the list of actual benefits is growing and that of potential benefits is long and open-ended. The one actual example of hazard he cites was an attempt to insert a gene from a brazil nut into a soya bean; the project was abandoned at an early stage of development over ten years ago because of possible allergic responses to the brazil nut protein.

I came to the subject with an open mind, without prejudice or personal interest, either academic or industrial, for or against the technology. On reviewing the arguments on both sides, I have reached the firm conclusion that the evidence overwhelmingly supports the case for promoting the genetic modification of plants. My conclusion is based on a number of authoritative, peer-reviewed, articles in respected scientific journals, reports by leading scientific institutions from different parts of the world,[5] and two authoritative studies by the Nuffield Council on Bioethics. The first of these was a thorough study published in 1999 into the ethical and social implications of GM crops, conducted by a body that included representatives of consumers and Green groups as well as leading scientific experts. It concluded that there was a moral imperative for making GM crops readily and economically available to people in developing countries who want them. The second, set up in 2003 to re-examine this finding in the light of new evidence, strongly reaffirmed the conclusions of its predecessor.[6] Furthermore, there is a growing body of evidence of actual benefits from the cultivation of GM crops, particularly for small farmers in the developing world.

What surprised me most as I studied the evidence was the vehemence and extent of the opposition to genetic modification of any kind. The opposition has assumed all the characteristics of a religious crusade. It has become the main focus for all the suspicions currently held by environmentalists about recent developments in science and its opponents seek to stop its development altogether. Officially, Green lobbies demand a 5-year moratorium on the commercial cultivation of GM crops. In 1999, the Governments of the European Union did indeed declare a *de facto* moratorium, but the European Commission, which seeks to base its actions on evidence, has overruled attempts by regional authorities in various EU member states to declare their regions GM-free zones. The Green lobbies want the existing moratorium extended. It is clear that if their demands are conceded, they will ask for further extensions indefinitely. What gives this demand extra weight in Britain is that it comes not only from the usual suspects, Greenpeace, Friends of the Earth, the Soil Association, and their various offshoots, who adopt an essentially eco-fundamentalist stance against GM crops, but from some NGOs who do admirable work to alleviate hunger, poverty, and disease in the developing world. Why do they join Greenpeace and its allies in denouncing a technology which major academies of science from both the developed and developing world support and which the independent experts from the Nuffield Council on Bioethics say we have a moral duty to make available to the developing world? Their support for the opposition suggests the case against GMOs should be taken seriously and has conveyed the impression that those who care about the developing world should also join the opposition.

However, despite the high standing of these aid organizations, I find little merit in their arguments and even less in their approach. These two very disparate groups, the aid agencies and the Green lobbies, seem to have formed an esoteric club in which members only talk to each other, regarding the views of outsiders, however expert, as irrelevant. Indeed they have very little, if any, contact with experts in the field. Since the club consists of members who have high-minded motives such as feeding the hungry and saving the planet, they assume that what any one of them says must be right, so critical judgment is suspended. No regard is paid to the most elementary rules about respect for evidence.

These are serious allegations, but they are solidly based. For example, a widely quoted report 'GM crops—going against the grain' published in 2003 by ActionAid, a charity dedicated to the relief of poverty, argues that these crops cannot benefit the Third World.[7] It ignores the findings of independent experts, the Brazilian, Chinese, Indian, and Mexican Academies of Sciences, the Third World Academy of Sciences, the National Academy of Sciences, USA, and four separate reports by the Royal Society, as well as the two reports from the Nuffield Foundation published in 1999 and 2004. Instead its report quotes numerous references to studies by Greenpeace, Genewatch (an offshoot of Greenpeace), other Green lobbies, and its own branches, which, not surprisingly, confirm its adverse view of GM crops. Any attempt at an objective study would at least have referred to evidence on the other side, especially when it has such weighty scientific authority.

If a document of this kind had undergone the critical analysis to which political party manifestos are subjected at election time, it would have been torn to shreds. But because it is published by an aid agency and relies on reports by Green lobbies to which most newspapers are sympathetic, the press treats it with deference.

Its report of the Golden Rice project illustrates the general approach. Few projects address the problem of disease in the Third World more directly. To quote the Nuffield study:

In 1995 clinical Vitamin A Deficiency (VAD) affected some 14 million children under five, of whom some three million suffered xerophthalmia, the primary cause of childhood blindness. Two hundred and fifty million children had sub-clinical deficiency, greatly increasing their risk of contracting ordinary infectious diseases such as measles. In many developing countries such diseases contribute significantly to high mortality rates. At least one-third are found among poor people in Asia who rely on rice as their staple crop and for whom alternative sources of vitamin A are usually unaffordable.[8]

Rice has now been genetically modified to contain a precursor of vitamin A: a bacterial gene together with two genes from daffodils have been inserted into an edible strain of rice to make it synthesize the micronutrient β-carotene which is converted into vitamin A in the body. The project has been widely hailed as a development with enormous potential for good, though not as the 'silver bullet' that will suddenly cure all infant blindness or vitamin A deficiency in the developing world. Other measures are also needed, such as the fortification of food with vitamin A and education in nutrition.

The ActionAid report says, correctly, that it will take some years before the new crop can be grown, but does not mention that this delay is largely due to EU regulations passed as a result of pressure by the Green lobbies. It then argues that the Golden Rice project is worthless and cites as evidence a purported finding by Greenpeace (who else?) that 'A child would have to eat about seven kilograms a day of cooked Golden Rice [equivalent to 3 kilograms of uncooked rice] to obtain the required amount of vitamin A'.[9] The Nuffield report (a group of experts who treat evidence with circumspection) quote Potrykus and his colleagues, the researchers who originally developed Golden Rice: their estimate, which relies to a large extent on data from the Indian Council for Medical Research, is that daily consumption would need to be equivalent to 200 grams of uncooked rice, that is *one-fifteenth* of the amount claimed by Greenpeace. Further research suggests that the amount needed will be substantially less.[10] Among other mistakes in the Greenpeace report is the assumption that golden rice has to supply *all* the vitamin A a child needs. Most suffer from a *deficiency*, not a total lack.

A balanced treatment would at least have referred to the estimate of the original researchers, a source that is *prima facie* more authoritative than Greenpeace. To ridicule the enormous potential of Golden Rice, as ActionAid does, and dismiss it on the basis of claims by Greenpeace with its record of extreme partisanship is like quoting the Pope as an unbiased authority on contraception. (For further comment on Greenpeace's attitude to Golden Rice, see Chapter 5, p. 130). ActionAid is an organization that does valuable work in developing countries, but its document on GM crops is not worth the paper it is written on and the treatment of the Golden Rice project is nothing short of scandalous.

The same lack of intellectual rigour was evident in a report 'Feeding or Fooling the World—Can GM crops really feed the hungry?' circulated in 2002 by a group called the Genetic Engineering Alliance, a coalition of 120 UK-based organizations including ActionAid and other leading aid agencies, advocating a minimum five-year moratorium on the cultivation of genetically modified plants.[11] It shares many of the characteristics of 'GM crops—going against the grain'. Every possible quotation that supports, or might appear to support, the case for a freeze is cited, irrespective of its academic worth; no evidence against is mentioned, however eminent and independent the source. Pronouncements made by extreme eco-fundamentalists such as a propagandist called

Vandana Shiva, whose fanatical opposition to genetic engineering makes the Soil Association look like the model of reason and moderation, are treated as self-evident truths. Indeed, in places this report seems almost deliberately designed to mislead.

For example, one of the authorities most frequently cited is Gordon Conway, President of the Rockefeller Foundation. He is quoted no less than twelve times on a variety of topics including the dangers of introducing new technologies too soon, the need to remedy inequalities in order to make agricultural reform effective, and the exaggerated claims made by industry and the media about the early introduction of Golden Rice. While every quotation is accurate, the implication is clear: Conway and the Rockefeller Foundation support the call for a moratorium and have deep reservations about the cultivation of GM crops. There is no hint that the Foundation is a leading promoter of the technology in the developing world and funded part of the research on Golden Rice. Conway himself has written many articles and made many speeches expounding (but not exaggerating) its potential and that of other GM crops. Again, the document cites Ismail Serageldin (Director of the Library of Alexandria), the World Bank and the World Health Organization on various issues, implying that they support a moratorium, when in fact all of them support the development of GM crops. Once more, a balanced presentation would have mentioned that these eminent authorities oppose a moratorium.

These are only some of many examples that demonstrate the deeply biased nature of the opposition to genetic modification. Some of its opponents may be trying to make their case by citing evidence, but the hard core of the opposition is not concerned with evidence. If it were, activists would not ignore evidence, however weighty, that contradicts their argument, and quote only evidence, however flimsy, that confirms their prejudices. Their real objections are ideological and based on dogma. Unfortunately, this dogmatism has infected the reporting of developments in agricultural biotechnology by many sections of the media and has helped to turn public opinion against the science.

The British public, after all, did not come to its present mood of antipathy to GM crops unaided. It showed no instinctive hostility to GM products. A genetically modified tomato puree, clearly marked as genetically modified, sold extremely well until 1999, when the press mounted a campaign against 'Frankenstein foods', sparked off by wide publicity given to a research study on the adverse effect of GM potatoes on rats. A Dr Arpad Pusztai had claimed that potatoes genetically modified to contain a lectin gene from a snowdrop affected the immune system of rats that ate them and could be harmful to human beings.[12] The research itself was discredited—the study was heavily criticized by the Royal Society, which said the work was flawed in many aspects of its design, execution, and analysis and that no conclusions should be drawn from it. The Royal Society stressed how important it was that research scientists should expose new research results to others able to offer informed criticism through the process of peer review before releasing them to the public arena.[13] This has not stopped nearly all sections of the media—urged on by Greenpeace, Friends of the Earth, and their allies—continuing to stress health hazards from GM crops. No wonder the public does not want to eat GM food. Where else can it obtain its information about scientific developments except through the media? The lobbies calling for a ban on GM crops now cite public hostility, which they have themselves helped to create, as one of the principal justifications for imposing one.

How does GM methodology differ from traditional plant breeding?

To understand what has caused the furore, the first question to answer is how, if at all, GM technology differs from traditional plant breeding. Most people, however limited their knowledge of biology, know that plants and animals consist of millions of cells. Each cell contains thousands of genes (human cells have about thirty thousand) composed of DNA, which contain the instructions that control the development of the organism. Small, naturally occurring, changes (mutations) are always happening to these genes. We now know that this is because the information residing in them can be affected by a variety of external influences (such as tobacco smoke or irradiation) or by spontaneous accidents that cause minor and occasionally major changes in the behaviour of the cells and thus in some characteristics of the plant or animal. When a change enables it to adapt better to its environment, it has a selective advantage over its competitors.

Since the discovery of the genetic code, we have learnt more about what controls the process of growth. It has become possible to take a single gene from one plant and introduce it into another, or to switch on or off a particular gene in a plant and alter that plant's development. This is the essence of genetic modification. We have discovered that the genetic code is almost identical for all kinds of living things. We can therefore transfer genes between different species of plant and also between bacteria, animals, and plants. It is worth noting in passing that the same technique is used to make drugs for the treatment of certain diseases, without incurring the wrath of green pressure groups.

In essence, GM techniques are more precise and quicker than traditional methods of plant breeding, rather like using a laser beam instead of a hacksaw. Traditionally plants with desirable characteristics have been crossbred with other plants, but it is impossible to forecast the consequences accurately. Small genetic changes on which traditional breeding depends are random events and it is likely that not only sought-after genes will be transferred to a new hybrid, but also many others, some of which may have undesirable effects. It may take generations of repeated backcrossing to eliminate unwanted traits. This is not only chancy and time-consuming but expensive.

Some techniques used traditionally to produce mutations might cause the public some concern if they knew more about them. For example, for over fifty years, long before modern biotechnology was developed, pest-resistant crops were bred from seeds and plants that had been artificially bombarded with gamma rays to alter their DNA. The idea that food may be irradiated has in the past provoked public reaction as hostile as that against genetic modification (with as little evidence that it causes harm to human health). Yet irradiation of seeds has proved an effective way of changing DNA, causing mutations, some of which can then be selected, for instance to yield more pest-resistant plants; over 2000 varieties of seeds that have been irradiated to produce desired mutations are in regular use. Golden Promise, the most successful malting barley in the 1970s and 1980s, was a mutant produced by gamma irradiation of barley seeds. Ironically, in their desire to avoid synthetic chemicals, organic farmers have been more dependent than other farmers on crop varieties generated by irradiation. (If the public were aware of this, enthusiasm for organic food might begin to wane.) Like the modern plant biologist who uses laboratory methods of genetic modification, the traditional plant breeder generates new combinations of genes but may add some tens of thousands of genes to the mix, making the result unpredictable. Irradiation alters both chromosome structure and genome sequence in a way that is totally random; genetic engineering, by contrast, introduces a single gene, or sometimes a few genes, into a settled genetic background.[14]

Supporters of genetic engineering sometimes argue that opposition to it is misguided because it is not essentially new, just more efficient. After all, throughout the ages since farming began, farmers have sought to improve on nature, breeding cows that produce more milk, sheep that give more wool and a variety of crops with higher yields. Domestic pets have been selectively bred for countless generations, and over the course of time some varieties have changed out of all recognition. What is the Pekinese but a specially bred, some might say monstrous, descendant of the wolf? Genetic modification in the laboratory, it is argued, is not doing anything that has not been done before, but is doing it more quickly, more successfully, and more selectively.

While there is much force in this argument, it does not answer the misgivings of most critics. Genetic engineering is clearly a much more powerful tool than we have had before, so much so that people feel quite irrationally that there is a difference in kind, not degree. It is part of a new power of science that makes some people feel science is out of control. The genetic modification of plants is often mixed up in their minds with the cloning of human beings. Well-founded ethical objections to the latter are cited as reasons for opposing the deliberate alteration of genes in plants. Opponents of GM crops frequently make a scary link with BSE. In Switzerland, anti-GM campaigners launched the slogan 'Mad cow disease equals Genetic engineering equals Catastrophe.' Similar announcements appeared in the German media. In Britain too, at the height of the hysteria about Frankenstein food in 1999, food writers frequently linked BSE and GM food. The possibility that there is any connection between the two is so remote from scientific reality that the only sensible response is one of despair.

A balanced judgement

In the current debate about GM crops, national and international academies of science, representing leading biologists from all over the world, expert in the theory and practice of genetic modification, are ranged on one side and Green lobbies and their allies, which include most of the press, are ranged on the other. The former argue on the basis of evidence; the latter show little regard for evidence. The former support the technology in general but believe we should examine each case on its merits; the latter prefer generalizations to the examination of products case by case.

   On the basis of our present knowledge, it cannot be argued that no GM crop can ever do any kind of harm in any circumstances. It is difficult to name any technology that did not involve some risk initially and never had any adverse consequences: there is no reason why genetic modification should be any different. Nor will GM crops that are beneficial to the environment in some places be so everywhere: plant biologists invariably stress that the success (or failure) of each crop will depend on the particular circum-stances. Furthermore, as with any new technology, it may not achieve all the hopes placed in it. Nor is it claimed that GM crops alone will solve the problem of how to feed the world.

Subject to these provisos, I believe the evidence supports five basic propositions:

(i)   The likely growth in world population will require more efficient agriculture and better use of land.

(ii) Genetic modification of crops is one of the most promising ways to increase such efficiency.

(iii) The technology can help relieve poverty and feed the hungry.

(iv) It can be used to reverse the decline in biodiversity and improve the environment.

(v)   It can help reduce human disease.

If these propositions are true, scientists do not need to apologize for genetic modification. They should be shouting its virtues and potential benefits from the rooftops.

(i) The need for more efficient agriculture

Agriculture over the last fifty years has been a success story for most of the world. About two hundred years ago, the Reverend Thomas Malthus famously predicted that the production of food could never keep up with the growth of population.[15] Yet contrary to his predictions, the expansion of the human population, which has been far greater than he could ever have imagined, has not outstripped our capacity to produce more food. From 1950 to the early 1990s, the world population more than doubled (it actually increased 2.2 fold) while food production nearly trebled (actual increase of 2.7 fold).[16] Nevertheless, although the absolute as well as relative number of those who are undernourished dimin-ished, over 800 million people are still undernourished today. There are also many parts of the world that have seen little or no improvement in recent decades: in most of Africa, for example, food production has barely kept up with the growth of human population.

This increase in the efficiency of agriculture in the last fifty years has largely been due to the Green Revolution, and its father, Norman Borlaug, has probably saved more human lives than any other person in the twentieth century. Before the Green Revolution, about half the population in the less developed countries did not have enough to eat; today the proportion is about one in five. But, as is generally acknowledged, the Green Revolution is running out of steam. In the 1990s, improvements in crop yields were much less than in the previous two decades, and for a number of reasons such as shortages of water, loss of soil, new types of pests and diseases, and the fact that the world is running out of suitable land, the decline is likely to continue. Some crops such as semi-dwarfed rice and wheat that gave greater yields were planted in the best-suited lands. When the Green Revolution was launched, there was still spare agricultural land available; even so, its success was partly achieved by using up some of the world's resources. Although the extent of the loss of usable agricultural land has sometimes been exaggerated, there seems little doubt that over the past fifty years there has been considerable erosion and loss of the world's topsoil, of its agricultural land, and its forests—and many scientists claim that there has been a serious loss of biodiversity.[17]

Even without these losses, since there is no longer spare land available, prospects look decidedly grim unless we find a new source of productivity to make up for the fading effects of the Green Revolution. World population, as most forecasters agree, will increase from 6 billion people today to about 9 billion by 2050. To feed everyone adequately, we will need to produce at least 50–60 per cent more food per head.[18] It follows that food production will have to more than double—over 50 per cent more per head for some 50 per cent more people. But the population will also be much wealthier and more people will live in cities. We must therefore take into account the fact that eating habits will change and many more people will want to eat more fruit and vegetables and more meat. (They will also have more pets, mainly cats and dogs, whose demands cannot be overlooked, as they are unlikely to become vegetarians.) The best guess on the evidence before us is that in the next forty to fifty years we will need to produce three times as much food as we do today. This huge increase must be achieved, unlike fifty years ago, without the use of extra land. Indeed, poor farmers are already resorting to 'slash and burn' to maintain their livelihood and large tracts of forest are being destroyed. I have already mentioned the destruction wrought to tropical forests by Mexican farmers.[19]

The constant refrain from Green lobbies that we do not need new technology because there is enough food in the world is wrong and dangerously so. It is true that there may be food surpluses in the United States and Europe from time to time, but given the fact that bad harvests may occur, as they have in the past, and the uncertainty of agricultural policy in the future, we cannot assume that there will always be surpluses. In any case, they do not solve the problems of the present (otherwise there would not be hundreds of millions of people who are undernourished) and are irrelevant to the problems of the future. Campaigners ignore the unwillingness of Western governments to promote fair redistribution, its huge cost, the problems of meeting local dietary needs and preferences, and above all the devastating effect reliance on food aid from rich countries would have on the livelihood of local farmers in poor countries. The difficulties of reforming the Common Agricultural Policy in Europe and the increase in farming subsidies that came into force in the United States after George W. Bush came to power demonstrate how unwise it is to rely on rich nations basing their food policies on the needs of the developing world.

Furthermore, those who expect present surpluses to meet future needs seem blind to the scale of change that is needed. No one can argue that we are able to feed the present population of the world using farming methods of fifty years ago. Why should it be supposed that we shall be able to meet the huge extra demand of the next fifty years by today's methods? If, as the Green lobbies urge, we turn against science and reject the gains in productivity that biotechnology has to offer, the Reverend Malthus, Paul Ehrlich,[20] and other doomsters might even be proved right.

My first proposition, that the world requires more efficient agriculture and better use of land, is therefore clearly made out. It is scarcely credible that Green lobbies, who normally adopt a cataclysmic view of the future, can take their own forecasts of food surpluses seriously. Is the reason for this uncharacteristic complacency perhaps that it suits their obsessive and dogmatic opposition to GM crops?

(ii) Genetic modification improves agricultural efficiency

Because the new technology is quicker and more precise, it should, in principle, be possible to develop new varieties of crops more cheaply, with higher yields and better nutritional qualities. There is every reason to expect that a number of superior varieties will be bred and marketed in due course. However, normally a new technology will take time to be widely adopted and prove its worth. What is remarkable, yet has not been widely commented on, is how quickly GM technology has been adopted and how quickly it has produced evidence of benefit. The strongest testimony to its contribution to agricultural efficiency is that the speed and extent of the uptake of GM crops has surprised even the originators of the technology and its promoters in the biotechnology industry. By 2003, over 7 million farmers (the majority of them small farmers) in 18 countries were cultivating GM crops on 68 million hectares.[21] It is estimated that 80 per cent of the soya bean crop, 70 per cent of cotton, and 38 per cent of maize in the United States are now genetically modified. Over 4 million small farmers in China and many thousands in South Africa grow GM cotton. GM cotton saved the cotton crop of Australia when it was on the verge of collapse from the build-up of pesticide resistance. GM cotton is now being grown in India and Brazil and constitutes one-third of the total cotton crop in Mexico. Such rapid transformation is unprecedented, even by comparison with the uptake of new varieties of crops during the Green Revolution.[22]

Apart from the compelling evidence of their rapid spread, there is specific evidence that those GM crops already planted improve economic returns and yields. Figures from the American National Center for Food and Agriculture Policy in 2003 show that the six GM crops currently on the market—rape, maize, cotton, soya beans, squash, and papaya—produced an additional 1.8 million tons of food and fibre on the same acreage, improved farmers' income by $1.5 billion, and reduced pesticide use by 21,000 tons. Most of this success was due to the modification of soya beans to tolerate the herbicide glyphosate and of maize to express a *Bt* protein derived from the bacterium *Bacillus thuringiensis* that protects it from the European corn borer.[23] Evidence of improved efficiency also comes from the widespread planting of *Bt* cotton in the developing world. This evidence is of particular importance because it demonstrates the actual benefits from the new technology already achieved for small farmers.

B.thuringiensis is a naturally occurring soil bacterium. There are a number of different strains of the bacterium, the spores of which contain different crystal proteins that are toxic to the different insects that ingest them. Each protein kills only one or two kinds of insect, that is, its action is highly selective. *Bt* cotton has been made resistant to its common pest, the boll weevil. China has developed several different varieties of *Bt* cotton, cultivation of which has resulted in a drop of between 60 and 80 per cent in the average application of pesticides. The crop has brought substantial financial savings, as much as $500 per hectare; the expansion of production is bringing prices down, so that the benefit of greater efficiency is now being shared between farmers and consumers.[24]

In KwaZulu, South Africa, 92 per cent of cotton growers had adopted *Bt* cotton over the five-year period to 2003, and the number is still rising. By their second season their income had increased by 77 per cent, despite the seeds being twice as expensive as those for non-modified cotton. All these benefits went to small farmers, most of them women, and in general, the smaller the farm, the greater the benefit. In Mexico, for the years 1997 and 1998 combined, an estimated $6 million surplus was generated by cotton farming, of which 86 per cent went to the farmers. Globally the benefits from GM cotton for the period 1998–2001 were estimated to be $1.7 billion.[25]

In 2002, India also licensed the cultivation of *Bt* cotton, in the face of strong opposition from Green NGOs. The Indian President A. P. J. Kalam, in explaining the decision to grant licences, urgently recommended biotechnology for agriculture 'to launch a second green revolution.' He said: 'All our agricultural scientists and technologists have to work for doubling the productivity of the available land ... 400 million Indians are struggling to come out of poverty. ... Technology is the only tool we have'.[26] One reason for the change in the policy of the Indian government was that those fields which had been planted with *Bt* cotton illegally in 2001, without government approval, survived a major infestation of pink bollworm virtually unscathed, while conventional cotton plants were devastated. Field farm trials of *Bt* cotton, in years with high bollworm incidence, gave gains in yield of 80 per cent compared with those from conventional varieties.[27] The cultivation of *Bt* cotton in India is spreading rapidly. In 2003 the area on which GM cotton was being cultivated was 100,000 hectares compared with 30,000 hectares in the previous year and according to the Agricultural Commissioner, C. D. Mayee, it was planned to license 10 further varieties of GM cotton in the following two years.[28]

The story of *Bt* cotton is exceptional in that actual benefits in the developing world have been realized so quickly. Other gains from biotechnology must still be regarded as potential rather than actual, but their list is long and their importance to the developing world is likely to be immense. Indeed the larger developing countries are investing substantial resources in plant biotechnology. China's public investment alone is now over $100 million a year and rising. China has taken the lead in GM technology for the developing world: it has developed 141 types of GM crops, 65 of which are already in field trials. By 1999 it had approved 30 environmental releases and 24 commercializations of GM crops, mostly targeted at insect pests and diseases. Its motto appears to be: 'Let a thousand GM crops bloom'.[29] Research institutes in India and Brazil also have excellent scientific capacity and are likely to become the source of many new crop varieties for small farmers.

Aid agencies stress the role that can be played by conventional plant breeding in producing new crops for the developing world. However, these take longer to develop than GM crops and certain traits such as *Bt* insect resistance cannot be inbred without transgenic technology. Thus many of Africa's problems, such as the need for drought tolerance, and resistance to the parasitic weed *Striga* or to maize stem borers, cannot be solved through conventional means. A way of controlling *Striga* by means of genetic modification has now proved successful in field trials.[30] Seeds of maize made resistant to the herbicide imazapyr are coated with the herbicide so that it leaches into the soil around the base of the plant and blocks the growth of the weed. While conventional plant breeding has achieved some improvement, especially for maize, there have been no such advances for the most important crops for the very poor, such as millet, sorghum, yams, and cocoyams.[31] *Bt* genes are now being introduced into maize and Belgian and East African scientists are working on inserting genes that will make bananas tolerant to Black Sigatoka and to nematodes and banana weevils, a development that could increase banana production by some 75 per cent. Field and greenhouse tests have demonstrated the effectiveness of *Bt* rice against pests and diseases in China, India, and Pakistan, and a GM rice resistant to bacterial blight is also being developed in a number of countries.[32] Yet a leading writer on agricultural matters who advertises his background in biology (and is a strong campaigner against GM crops) claimed in a book published in 2003 that GM technology has 'contributed nothing truly worthwhile to wheat, rice and maize *and is unlikely to do so*' (my italics).[33] GM crops may have no adverse effect on human health, but clearly opposition to GM crops can cause intellectual blindness.

Three other developments can be mentioned, chosen almost arbitrarily from a long list. The first is apomixis, a form of plant reproduction that by-passes sexual reproduction, thereby producing seeds that are genetically identical with the maternal parent. It occurs naturally in about 400 plant species distributed over more than forty plant families. Biotechnology can develop the process in crop plants, with the great advantage that it allows farmers to select seeds of plants with attractive characteristics and save them year after year for subsequent plantings. Cultivars of seeds for crops such as cassava, potatoes, sweet potatoes, and yams can be propagated from seeds free from pathogens. The process saves storage, shipping, and planting costs. In the future, apomixis could be one of the most important innovations in the history of agriculture, benefiting all farmers, including small farmers who have not shared in the benefits of previous technical revolutions.[34]

The second is the development of salt-tolerant crops and crops that can grow in arid regions of the world. Already a transgenic tomato has been engineered that can thrive on salty water, and cereals that can withstand drought and salinity are undergoing laboratory tests. If salt-tolerant crops can be grown commercially, huge tracts of land that have become infertile through irrigation can be brought back into cultivation. Irrigation has been a cornerstone of agriculture for centuries, and played a vital part in the Green Revolution, but it has not done so without harmful consequences. As water used for irrigation evaporates it leaves traces of salt that gradually accumulate. Every year 25 million acres are lost to salinity, which has now affected 40 per cent of the world's irrigated land and 25 per cent of the United States. The transgenic tomato is not only able to grow in saline soil, but its leaves take up 6–7 per cent of their weight in sodium chloride and thus actually remove salt from the soil.[35] A gene that enables a plant to survive prolonged periods without water in desert conditions has also been inserted into rice. This creates rice that produces a sugar that protects the plant during periods of dehydration and allows it to survive periods of drought. It is hoped to introduce the trait into other crops, such as maize, wheat, and millet.[36] Since one-third of the world's arable land is affected by drought, the potential benefits are self-evident.

The third example may seem relatively minor by comparison, but is worth mentioning to illustrate the important contribution that is still being made by publicly financed research in Britain. It is a project aimed at protecting potatoes in Bolivia against damage by nematodes, a small kind of parasitic worm. The rural population there, one of the poorest countries in the world, depends mainly on the potato as its staple food, but because of losses caused by nematodes the area of smallholdings required for its cultivation is twice what it need be. Past efforts to improve productivity by about 2.1 per cent a year have merely kept pace with population growth. There is no more land available to increase agricultural holdings, except by destruction of the wilderness. Scientists at Leeds University are developing a simple, cost-effective way of controlling nematodes by genetic modification, using the gene for cystatin, a protein that is present in various seeds, including those of rice, maize, and sunflower that is toxic to the nematodes. The protein is also present in human saliva, hence it is unlikely to pose a risk to human beings. Like other toxic proteins incorporated into GM plants, its very specific nature ensures that any risk to biodiversity is minimal, less than that from conventional methods of pest control, and the chance of such GM potatoes transferring their pest-resistant qualities to the wild is negligible. The benefits will be enormous, not only from increased potato production, but also from the release of land for cultivation of other crops which could greatly improve the diet of the rural population. Bolivia has one of the lowest calorie intakes per capita in the world.[37]

It is not argued that all of these examples of potential benefit will necessarily translate into actual benefit; nor is it denied that, even if they do, it will take time before the actual crops are grown. Nor do those who wish to promote them argue that they are the only means of increasing crop yields or that GM crops alone will feed the world, a simplistic claim often referred to by the opponents of the biotechnology industry but not made by scientists themselves. In many parts of Africa, subsistence farmers are too poor to afford fertilizers and modern machinery and they would benefit greatly from improved traditional farming methods or other techniques that do not involve the use of biotechnology. Often higher productivity depends on improving the infrastructure of a nation, for instance by building better roads, or on profound political changes. But from these obvious facts, some activists go on to argue that the world does not need genetically modified crops—and should not grow them—as if the prospect of growing crops on arid or saline soil, for instance, can be dismissed as an irrelevance. It is like arguing that there is no need to do further medical research or develop new life-saving drugs, because some people's ill health is due to poverty. The same opponent of GM who declared that it had no likely benefits to offer to the improvement of rice, maize, or wheat crops, dismisses the present 'vogue' for GMOs as 'a South Sea Bubble'.[38]

My second proposition, that genetic modification is one of the most promising ways to achieve greater agricultural efficiency, is therefore supported by a formidable body of evidence.

(iii) GM crops can help to relieve poverty and hunger

Improving agricultural efficiency will itself help to relieve poverty and hunger. Two-thirds of the world's poor depend mainly on agriculture for their livelihood and 70 per cent of them live in rural areas. The poor are not only farmers and those employed by farmers, but also those employed in non-farming activities dependent on farming.[39] Higher yields from small-scale farming are of key importance and GM crops are particularly suited to increasing yields for small farmers because they reduce crop loss caused by pests and plant diseases. If it could eliminate these losses, the technology would double food production in Africa. It is not a technology, as its opponents constantly allege, whose benefits depend upon the spread of monoculture.

Again, it is important not to overstate claims made for bio-technology or to generalize. How it is applied is of central import-ance. Evidence clearly shows that Bt cotton has benefited millions of small farmers, mainly in China but also increasingly in India and South Africa. Greater yields from sweet potatoes in Kenya and nematode-resistant potatoes in Bolivia and Peru will help the poorest of the poor. On the other hand, herbicide-resistant crops which improve productivity and encourage farmers to displace labour are unlikely to be suitable for most developing countries, except perhaps in places like Kenya, where AIDS has caused a shortage of labour and children have been forced to work in the fields instead of going to school.[40] However, in general, by reducing or even eliminating the need for pesticides, and potentially by making plants drought- or salt-resistant, genetic modification is better designed to help small farmers than was the Green Revolution. Although it saved millions from hunger, increased employment and improved living standards for landless labourers, the Green Revolution depended heavily on greater use of fertilizers, which some could not afford, and on irrigation, which was not always available, especially in Africa. In some places it also caused major changes in the habits of local communities. By contrast, according to a leading African biologist, genetic modification provides 'its packaged technology in the seed, which ensures technology benefit without changing local cultural practices'.[41]

From these examples it is clear that my third proposition, that genetic engineering can help reduce poverty and hunger, is also solidly based.

(iv) Genetic modification can be used to increase biodiversity and improve the environment

Some people may regard this as the most controversial of my five propositions. While there is general agreement among leading scientists—indeed it is a view supported by every authoritative body that has examined the evidence—that there is no reason to expect GM crops to be a danger to health, many plant biologists do not dismiss the possibility of damage to biodiversity and to the environment, although of course they will add that it depends on the particular crop, how and where it is grown, and the particular circumstances: scientists, unlike Green lobbyists, avoid generalizations. (For a discusssion of the possible dangers to biodiversity and also of the Field Scale Evaluation Trials in the UK, see next chapter.)

However, whereas GM crops were not primarily developed to produce environmental benefits, one of their advantages is that they can be used to reverse the harmful effects of intensive farming. Modern intensive farming has greatly improved productivity and the quality of food, but it has converted forests and other natural habitats of wildlife to agricultural use, diminished the variety of crops and has encouraged monoculture. In Britain it has reduced the number of hedgerows, which are an important habitat for wildlife. Pesticides and herbicides have been so successful in eliminating pests and weeds that they have affected the insect population, which in turn has affected other animals, including birds, which feed on insects that normally live among the crops and weeds.

Since GM technology can improve productivity, it frees farmland which can then be set aside for new wilderness. Through the more efficient use of land, it can avoid further deforestation. Experience to date shows that in the United States the widespread introduction of herbicide-tolerant crops has cut the use of herbicides, allowing more weeds to grow for longer, thereby preserving insect populations. One field trial performed at Broom's Barn, Suffolk, in Britain showed that herbicide-tolerant sugar beet could be cultivated in a way that allowed weeds to grow for longer than in fields where beet is farmed conventionally. The surviving insect population was eight times larger than in the area farmed under the control regime. The potential effect on bird life is likely to be highly beneficial.[42]

Indeed one of the main ways in which pest-resistant GM crops benefit the environment is their beneficial effect on the non-target insect population: they reduce the use of pesticides that will kill all insects, not just the pests of that particular crop. In 1997, Bt crops saved $2.7 billion of the $8.1 billion spent annually on pesticides in the United States. In Australia, Bt cotton not only saved the cotton industry, but reduced spraying insecticides by 60 per cent. In China the use of formulated insecticide in 2001 was reduced by nearly 80,000 tonnes.[43]

Another case of a GM crop that improves the environment is the variety of *Bt* maize that contains the toxic protein that specifically kills the corn rootworm, the larva of a beetle that causes huge destruction of the maize crop in the US Midwest. Previously the pest was controlled by drenching the soil with enough of a potent pesticide to kill the rootworm, but this caused serious damage to beneficial organisms in the soil. Thus, the *Bt* toxic protein not only protects the maize, but also the soil.

A second major environmental benefit, already proved in practice, is the impact of herbicide-tolerant crops on the need to plough as mentioned earlier. Traditionally, ploughing has been synonymous with farming. 'Arable' is derived from the Latin *arare*, 'to plough', and arable land is by definition land that can be ploughed, as opposed to pasture or woodland. No crops could be grown in the past without the plough because weeds would take over. But ploughing is not environmentally friendly: it damages the soil. Unploughed soil is rich in organic matter that provides food for plants, as well as being rich in earthworms, insects, and microbes. Ploughing stirs up the soil, turns it over and over and makes it homogeneous and lifeless. It breaks up birds' nests and drives away small mammals. It causes run-off of nitrogen and phosphorus that pollutes rivers and lakes and blights aquatic habitats. It also causes soil erosion, because it loosens topsoil that can be washed away by rain and blown away by wind. That is one reason why the expansion of agriculture in the last few decades has led to the loss of a significant proportion of the world's topsoil. Ploughing also releases greenhouse gases, such as carbon dioxide that are stored in the soil.

Herbicide-tolerant crops can make ploughing obsolete. Most have been engineered to tolerate the broad spectrum herbicide glyphosate, produced by Monsanto, which is less toxic than other herbicides and needs to be applied less frequently. Most importantly, when used on herbicide-tolerant crops there is no longer the same need to plough to control weeds. More than a third of the soya bean crop grown in the United States is now grown in unploughed fields. A study published by the Conservation Technology Information Center in the United States reported that no-till farming had increased by 35 per cent since herbicide-tolerant crops were introduced. This has led to major savings of many kinds: nearly one billion tons of soil per year are no longer being eroded, $3.5 billion savings are projected in the cost of dredging rivers, cleaning road ditches, and treating drinking water, and 3.9 gallons of fuel per acre can be saved.[44] Researchers from Michigan State University reported that no-till farming could reduce the impact of modern farming on global warming by 88 per cent.[45] It is nothing less than an agricultural revolution. It changes the way farmers have treated the soil since time immemorial. If the Soil Association and other organizations, whose declared aim is 'to feed the soil', mean what they say, they would abandon the tractor and plough and become leading champions of genetic modification.

My fourth proposition is therefore also firmly based on evidence.

(v) GM technology can help reduce human disease

Cultivation of GM plants is already having a dramatic effect on the health of farm workers who are now less exposed to toxic pesticides. In China, where farmers traditionally used back-pack sprayers and did not wear any protective clothing, pesticide poisoning was high before they grew *Bt* cotton; it has now declined by a third. In South Africa, where *Bt* cotton is now sprayed twice a season instead of the eight times or more required by unmodified cotton, local hospitals have reported a reduction from 150 to a dozen in the number of cases of burns and sickness due to agricultural chemicals.[46] Similar results have been reported from Mexico.

But GM technology has a potentially greater contribution to make to the prevention of disease. The genes of plants can be very effectively added to for medical purposes. Biotechnology was first developed with new treatments for disease as one of its principal objectives and an increasing number of medical products, including vaccines, are now made by genetic engineering of bacteria and yeasts. For example, human growth hormone obtained from bacteria engineered to contain the human gene has now replaced extracts from pituitary glands of cadavers, that sometimes contained the agent that causes CJD. The clotting factor, Factor VIII, previously obtained from blood donors, is another product of genetic modification; it is a safer source of treatment for people with haemophilia since it is free from contamination with the HIV virus that causes AIDS. Other medical applications of genetic technology include the manufacture of several vaccines against hepatitis, of erythropoietin that stimulates the production of red blood cells, and of monoclonal antibodies to treat cancer. Whether biotechnology is used for medical or agricultural purposes, the technology is the same.

Apart from the Golden Rice project developed to overcome vitamin A deficiency, rice can be enriched with iron by genetic modification, to help about 400 million women in the world who are anaemic as a result of iron deficiency and who tend to produce stillborn or underweight children and are more likely to die in childbirth. Anaemia in Asia and Africa is a factor that contributes to about 20 per cent of maternal deaths after childbirth.[47] More contributions to health will come from the development of crops with greater nutritional value. India, for example, has already developed a potato genetically modified to produce increased protein and has a six-year plan to develop GM crops that will provide a higher nutritional content.[48]

Another medical benefit of enormous potential significance that will be realized in the foreseeable future is the production of vaccines and pharmaceutical proteins in plants. Potatoes, bananas, and tomatoes are being modified to produce vaccines against diarrhoeal diseases, hepatitis B, and cholera and perhaps the most reliable form of delivery of oral vaccine will be in the form of food-based tablets.[49] Although vaccines are already available for many diseases that ravage the developing world, as the world's leading Academies of Science have pointed out they are expensive to produce and often difficult to store and use and they have to be administered by trained specialists. In some countries even the cost of needles is prohibitive. It is not surprising that vaccines reach only a small proportion of those people in the developing world who need them. The new vaccines will be swallowed, avoiding the need for syringes and needles.[50] They will not only be cheap to make and buy, but easier to administer and store; their expression in seeds will be a great advance, as mankind has safely stored seeds since agriculture began. They could in time prevent, perhaps even eliminate, some of the most devastating diseases on the planet. In the developed world too there is a long list of diseases that are likely to benefit from the production of pharmaceuticals in plants.

It is clear that the potential medical applications alone justify the conclusion of the first report of the Nuffield Council on Bioethics that there is a moral imperative to make the products of GM technology available to the developing world. Looking at the overall picture, given the actual proved benefits of the technology and its vast and expanding list of potential benefits, it is hard to see how any reasonable person can argue that it should be banned. Indeed the campaign against GM crops, in the harm it can do to the developing world, in some ways resembles Western hysteria about DDT, which has caused millions of deaths from malaria that could have been prevented (see Chapter 7). Of course, the evidence of benefits, however strong, does not prove that the technology cannot ever have adverse effects and does not carry any risks. The arguments against must therefore be examined in detail; but they will have to be exceptionally compelling to outweigh the benefits.

5

The Case against GM Crops

The human understanding when it has once adopted an opinion draws all things else to support and agree with it. And though there be a greater number and weight of instances to be found on the other side, yet these it either neglects and despises, or else by some distinction sets aside and rejects, in order that by this great and pernicious pre-determination the authority of its former conclusion may remain inviolable.

Sir Francis Bacon

See skulking Truth to her old cavern fled
Mountains of Casuistry heap'd o'er her head

Alexander Pope

THE case made against the development of GM crops is that they are a danger to health, that they damage the environment, that there is no need for them, that they cannot lessen poverty, hunger, and disease and, lastly, that their main, if not sole, purpose is to maximize the profits of multinational companies. First, however, there is one other issue that needs to be dealt with separately: many people are instinctively hostile to biotechnology because they feel we are 'playing God with Nature', or to put it in more general terms: 'any attempt to take liberties with nature is likely to backfire'.[1]

'Genetic modification is unnatural'

If the view that we must not 'tinker with nature' is held on grounds of religious dogma, it is difficult to answer with rational argument.

My instant reaction is to laugh it out of court and ask if it is unnatural to build aeroplanes since, if God had meant us to fly, he would have given us wings. Leaving dogma aside, it is hard to see what actions are involved that contradict holy writ or some other canon of hidden laws of the universe. By breeding new crops and animals, mankind has clearly 'interfered with nature' over the ages. Why, instead of crossbreeding different strains conventionally, is it unnatural, or usurping God's role, to take a single gene out of one plant and insert it into another? Why is this more unnatural than irradiating seeds to cause mutations? The philosophical distinction between 'unnatural' gene transfer and 'natural' induction of mutation by irradiation is obscure.

Many people who worry about tinkering with nature nevertheless accept the use of biotechnology for medical purposes, for instance by inserting human genes into bacteria for the production of human insulin. If so, why is it 'unnatural' to use the same technology to transfer genes from one plant to another? Is medical treatment good but improving the yield of plants to feed people bad? Is it wrong to kill pests more efficiently? It makes no sense to argue that the technology that makes plants resistant to harmful insects, fungi, or viruses should be rejected, but when it is used to make better drugs to protect us from life-threatening diseases it is to be welcomed. In many parts of the world, producing enough food is just as important for saving lives as supplying medicines.

It is often argued that what is unnatural is crossing the species barrier. But transferring the human gene that codes for insulin into bacteria also crosses the species barrier. The objection seems to be based on the assumption that animals and plants are chemically different. Not so. Human beings share over 50 per cent of their genes with bananas. What, then, can be the grounds for objecting to the transfer of a desirable gene from a different species such as a bacterium into a plant? All sorts of transfers or transplants raise opposition when first mooted. In the early 1950s, a special commission was set up in Britain to decide whether it was ethical to graft corneas from corpses into the eyes of the living. It was decided that to do so would be unethical (one reason being that corpses cannot give their consent). Nowadays such transplants are almost universally regarded as acceptable. When the first heart was transplanted from a clinically dead person into someone who would otherwise die, many people thought it was a step too far. Today heart transplants from human to human are an accepted part of medicine, and pigs' valves are commonly implanted into patients suffering from heart disease. If the transplant of such tissues between species is ethical to save human life, there can surely be no rational objection to growing transgenic fruit that resist frost with the help of a gene from a bacterium or a fish (in fact, the transfer of a fish gene into a strawberry, an act that has given rise to much concern among opponents of genetic modification, has not yet been documented).

Of course, no one denies that there comes a point when serious ethical issues arise. If we can clone sheep, there is no technical reason why in time it should not be possible to clone a human being, and some doctors have already declared their willingness to do so. The UK Parliament decided that, whereas human embryos up to fourteen days old may be used for stem-cell research, the cloning of human beings should be explicitly outlawed; but there are reasonable concerns that some countries may in time take a more permissive view. Eventually we will no doubt learn how to eliminate inherited diseases such as haemophilia, for example, by genetic modification. If we succeed in eliminating inherited characteristics that are undesirable, in due course we will also be able to select for desirable ones and we will have to face the issue whether this should be allowed. The choice will be ours: it will not be left only to scientists.

'GM crops are a threat to human health'

Polls suggest that the fear that eating GM food will poison us and make us ill is the main reason why people do not want to buy it. Initially they were happy to eat paste made from genetically modified tomatoes, clearly labelled as such, but scare stories led to the retailers withdrawing the product from their shelves. Is there any cause to be scared?

To begin with, there is no intrinsic or theoretical reason why GM crops should constitute a special risk. Every mouthful of meat, fruit, or vegetable contains DNA and the proteins it codes for, or their components. The DNA from transgenic crops is made up of exactly the same chemical components as all other DNA. Our digestive systems have evolved to break down the molecules so that they can be absorbed and used by our bodies. All GM crops are rigorously tested for toxins and allergenic proteins, using tests on animals as well as analytical tests in laboratories, before the crop is licensed for use. In this respect GM food is safer than conventional food, which has not been tested in the same way.

Next, 280 million Americans have been eating GM maize and food made from it, and GM soya beans and their products, for over seven years without any known damage to health. Foods containing products from GM crops are a ubiquitous component of the average American diet. Corn syrup and soya bean oil derived from GM crops together account for 70 per cent of processed foods on grocery shelves and GM corn and soya bean meal are present in the feed of livestock reared for human consumption. Most Americans therefore eat foods derived from GM crops every day of the year.[2] So, for that matter, do Europeans, since some 36 million tonnes of soya and soya meal are imported every year, some two-thirds of which are genetically modified and incorporated into animal feed. What has been less widely reported is the absence of litigation in the United States about the safety of GM food, and when lawyers in America can find no grounds to sue, there can be no basis for concern. In the past, unfounded scares, such as reports that electromagnetic fields cause cancer or that breast implants cause connective tissue diseases, led to numerous lawsuits (and the awards of massive damages); but not a single case has been brought alleging that GM crops harm human health, although the amount of GM food consumed increases substantially every year. GM crops are now grown worldwide on a massive scale as stated earlier. In 2003, the total acreage of transgenic crops under cultivation around the world was nearly 70 million hectares, an area more than twice the area of the United Kingdom. Still there is no evidence of harm to human health.

In 2003, the US Environmental Protection Agency (EPA) published an updated report on nine *Bt* crops, about current health and ecological data. It concluded that those crops pose no significant risk to the environment or to human health—and are less likely to be harmful than chemical alternatives.[3] Like the US Food and Drug Administration, the EPA is trusted by the American public, a fact that may contribute towards the clear differences in attitudes to biotechnology in America and Europe. Also significant was the response of the audience at a conference sponsored by the Organization for Economic Co-operation and Development (OECD) in Edinburgh in 2000, attended by some 400 world experts on plant biology as well as representatives from 'green' and other NGOs. The conference chairman asked whether anyone could produce any evidence of damage to health from GM crops. There was a long, telling silence. The conclusions of seven international academies of sciences have already been mentioned. In fact, there is scarcely a single plant biologist of repute who shares the apprehensions of the European public.

Nevertheless, this vast body of evidence has not convinced critics and they have decided to ignore it. Thus, in its report 'GM crops—going against the grain' published in 2003, as mentioned, ActionAid stressed 'the known risks to human health posed by GM crops'.[4]

Two particular concerns are often mentioned: the use of genes for antibiotic resistance as selective markers and the fear that GM crops may contain dangerous allergens. Initially, marker genes for antibiotic resistance were widely used in the laboratory to indicate whether the transfer of the desired new gene had been successful. What is the chance that environmental bacteria will acquire resistance to antibiotics from any crops that contain these genes? If we eat the plants containing these markers will the bacteria in our guts become resistant to antibiotics? Would this not destroy the effectiveness of the antibiotics in question? Many studies have now been done to determine whether a gene for antibiotic resistance in plants can survive passage through the intestinal tract to be acquired by bacteria there. No evidence of gene transfer has emerged.[5]

Once again, the scare about antibiotic resistance markers shows that Green activists are not interested in evidence. Friends of the Earth and one of their media supporters, John Vidal in *The Guardian*, claimed that serious health questions were raised by a study by scientists in Newcastle University, UK, because it showed that genetically modified DNA material from crops had found its way into bacteria in the human gut.[6] In fact, this was not the conclusion of the study. The workers summed up their findings by stating 'We conclude that gene transfer did *not* occur during the feeding experiment.' (my italics).[7]

In any case, the possible spread of antibiotic resistance from consumption of GM plants containing the markers is incomparably less likely than that from patients failing to complete their prescribed course of antibiotics. Nevertheless, to avoid the theoretical possibility of harm and since other markers are available, the seven National Academies of Science, as well as a further Royal Society report, recommended that the use of markers for antibiotic resistance should be phased out.

The danger that genes inserted into plant DNA might give rise to new allergens is also frequently cited as cause for concern. The issue arose partly as the result of a particular project to modify a soya bean to carry a gene from the Brazil nut, which has already been referred to in the previous chapter. The nut is a ready source of nutrients and energy and transfer of the gene could have made a valuable contribution to a more balanced diet for people in many parts of the developing world, at virtually no cost. Thus, in the mid 1980s, for the best of reasons, a small biotech firm in California explored the possibility. However, scientists pointed out that it was unwise to transfer a gene from a nut with known allergenic potential into food and the idea was dropped. Later, a firm called Pioneer Hi-Bred revived the project, this time for animal feed, and carried out tests to check whether the particular protein actually caused an allergic response in people. They found that it could, published the results, and no product was ever commercialized.[8] Subsequently Green pressure groups learnt about the gene from the Brazil nut and have continued to warn us ever since about the hazards of a product which was never marketed and whose dangerous properties were confirmed and publicized by the researchers themselves.

However, the potential presence of allergenic substances is clearly something we must guard against in all food products. For this reason, the seven Academies of Science, and more recently the Royal Society,[9] recommended that food regulators must closely scrutinize any novel proteins produced in plants, whether they are conventional or GM plants, which may become part of our food or of animal feed. It is worth adding that researchers in the US Department of Agriculture and the same biotechnology company Pioneer Hi-Bred have succeeded in genetically modifying soya beans to suppress the production of the protein that causes most allergic reactions to the beans themselves. In time they hope to knock out all soya bean allergens, to the lasting benefit of those who react to them.[10]

In the absence of evidence of actual harm to people, anti-GM activists resort to different tactics, saying instead that the crops have never actually been positively proved safe. One of the mantras of the anti-GM lobbies is that 'The absence of evidence of harm is not evidence of the absence of harm'.[11] Until the crops have been proved safe, they argue, we should ban them on the Precautionary Principle: 'Take no risks until we can be sure they don't exist'. It is of course impossible to prove a negative. No one can prove that 10,000 angels do not dance on the point of a pin. If absolute proof of safety were required before any food was passed fit to eat, we would all starve. All new crops are carefully regulated and have to pass rigorous scrutiny before they may be grown or sold in the shops. If anything, it could be argued that GM crops are over-regulated. In the UK alone, there are seven statutory or advisory government committees concerned solely with biotechnological issues and a further nine committees with an indirect biotechnological remit. Indeed GM crops have been tested for adverse effects far more thoroughly than conventional food and crops. Many long-established common food products have not been tested at all. If the potato were subject to the same tests as GM crops, it would never be licensed as safe. After all, it belongs to the same family as deadly nightshade and when potatoes grow green on exposure to light, they develop a range of dangerous toxins.

'Genetic modification harms the environment and reduces biodiversity'

Plant biologists take the possibility that GM crops could damage the environment seriously. Damage could take several forms: through the transformation of GM crops into so-called 'super-weeds' or by changing the nature of other plants through cross-pollination, often referred to as 'gene flow'. The crops might adversely affect animal life by reducing the number of insects, and consequently the number of birds and small mammals that feed on them. It is this possibility particularly that gives biologists some cause for concern.

These issues should be considered separately. A distinction should be drawn between the different possible effects of pest-resistant and herbicide-tolerant crops. In each case the test is not whether there is a risk to the environment from GM crops them-selves, but whether there is a greater risk from GM crops and their management systems than from the conventional crops and the usual system of administering herbicides or pesticides that they replace. If there is *some* risk from GM crops, but the risk is less than that from conventional crops, this is an argument in favour of GM crops. If the risk is no greater, that is not an argu-ment against them. Finally, in the case of gene flow, the issue is not whether it happens, because cross-pollination is a common phenomenon in nature and occurs between conventional crops,

but what harm can it do. Opponents of GM crops do not distinguish between the separate issues: if there is *any* reduction in
biodiversity or *any* evidence of risk to non-target species or *any*
gene flow from any particular transgenic crop, that, they argue,
proves the case against *all* GM crops, even if the net impact is
favourable.

Superweeds and irreversibility

Fear of 'superweeds' looms large in the pantheon of scare
scenarios. It is argued that if a transgene for herbicide- or pest-
resistance escapes from a modified crop into a weed or is transferred into a wild relative of the crop, the weed in question, or the
wild relative, will become resistant to herbicides or pests and flourish, thus creating a new breed of superweeds that cannot be controlled. Non-GM plants imported into the UK, for example
rhododendrons and Japanese knotweed, are apparently already
out of control. Perhaps we ain't seen nothing yet! Activists constantly repeat the refrain: 'once these genes have escaped,
they cannot be recalled' and argue that, once out of the bottle,
the genie (or gene) will cause lasting and irreversible danger to the
environment.

There are several reasons why these fears are misconceived. In
the first case, it is inherently unlikely that any GM crop itself can
become a weed. A plant needs at least twelve genetic traits to
become a successful weed, and domesticated crops are estimated
to have only six of them, thus placing them at a competitive disadvantage.[12] Weed populations, from which our crops were originally domesticated, are a sea of mutants; without such variation
they would not have been able to survive disease, predation, and
constant competition. Weeds found in fields and gardens are
simply tougher than agricultural crops, and this is as true for GM
crops as for conventional crops; those genes which have been
introduced in the laboratory do not convey any of the characteristics of weeds. Although the specially tough qualities of weeds
might be thought to make attractive candidates for transfer back
into crops by genetic modification, or indeed by traditional plant

breeding, they are not the sort of qualities that would enhance their performance as domesticated crops.[13]

It is not surprising, therefore, that a study of both GM and conventional crop plants introduced into the wild showed that agricultural crops did not survive. Four different crops (oilseed rape, potato, maize, and sugar beet) were grown in twelve different habitats and their fate was monitored over a period of ten years. In no case did transgenic plants persist longer than their conventional counterparts and neither kind survived for long.[14] The domesticated crops did not survive because they were not tough enough, just as our domestic pets would not survive for long if they were released into the jungle. There is therefore little danger of irreversible escapes. In any event, if GM crops did escape and survive, there are a variety of herbicides available to kill them. The only advantage wild relatives might gain over other plants from an accidental transfer of genes would be in arid or saline soil, if the genes had made them resistant to salt or drought, but any such risk would be outweighed by the benefits of growing crops on land where food cannot grow today.

There are lessons to be learnt from past introductions of new species. In 1988, in response to warnings about the irrevocable effects that release of genetically engineered organisms might have on the environment, the Office of Technology Assessment in the United States looked at the history of the introduction of new organisms. It found no special cause for concern. To quote its report:

The US Department of Agriculture's Plant Protection Office recorded over 500,000 introductions since 1898, mostly from outside the United States, including large numbers of plants, insects, and microbes. Although the proportion that has actually become established is not known, there have been occasional negative consequences, sometime severe or far-reaching. Few have been lasting and fundamental eco-processes and ecological relationships remain intact.[15]

In fact, when concern is expressed about new weeds, it should be noted that the introduction of different forms of flora and fauna into new environments has been a common phenomenon throughout the ages. In Britain it is generally thought that there are 1600 species of native flora and 3500 alien species, mainly introduced by horticulture.[16] Under the UK regulatory framework, one key test before a new product can be passed as suitable for release concerns its potential for producing superweeds: methods for suppressing resistant weeds must be proposed in any application for registration. The regulators believe in belt and braces, as is right and proper.

Gene flow and Cross-pollination

Green lobbies are much concerned about the danger of cross-pollination from GM plants to their wild relatives. There is no doubt that gene flow is difficult to avoid altogether. It occurs all the time. After all, the flow of genes to related species to produce new kinds of plants is one reason why such a wide variety of plants have evolved over millions of years. But the transfer of an introduced gene from GM crops to a wild relative could cause legitimate concern. For example, a variety of maize (*Teosinte*) grown in Mexico and its genetic diversity has made it the source of raw material for farmers and plant breeders everywhere who wish to create better strains of maize. There was therefore widespread consternation when it was claimed in a paper in *Nature* that genes from GM maize had crossed over into wild maize.[17] Although, in an unprecedented move, the editor later disavowed the paper because he was not satisfied with its quality, the possibility that such 'contamination' might occur cannot be ruled out.[18] It is not clear what damage would be caused, since the presence of an additional gene from GM maize would be unlikely to affect the unique qualities of wild Mexican maize; but it is obviously desirable that cross-pollination should be minimized, if not avoided altogether.

In the case of plants modified to manufacture pharmaceutical products like vaccines, gene flow, if not controlled, might have more serious consequences and must be avoided; research to assess such risks is clearly important. We do not want a new medicine in our cornflakes by mistake. In the field, proper separation distances between GM crops and their wild relatives must be established, to ensure that the amount of pollen that crosses the divide is so small that it can be safely ignored. There is no reason why this should not prove effective, as it certainly has in the past. Farmers have long learned to keep different varieties of the same crop separate. For example, rapeseed was originally grown in Canada to be used as an industrial lubricant because it has high levels of erucic acid, which can be harmful when eaten. Conventional plant breeders developed improved varieties of rapeseed—now called canola—with low levels of this harmful acid and canola is now a widely used cooking oil. Both types of rapeseed are still grown in Canada. Canadian farmers and processors manage routinely to keep the two varieties apart.

By-passing pollen

The best safeguards, however, would be to ensure that genes transferred into GM plants are not incorporated into the pollen or that modified plants are sterile. Apart from work on apomixis (see Chapter 4, p. 98), another promising technique has been recently developed that may reduce the risks of cross-pollination substantially. It should provide an elegant solution to many genuine concerns of environmentalists and could lead to new ways of using genetic modification for the benefit of mankind. This is the genetic engineering of the chloroplast genome.[19] Chloroplasts are separate entities within plant cells that have their own DNA, distinct from that in the nucleus. Transgenes with desirable characteristics have been successfully inserted into chloroplast DNA. The technical details are complex, but one result of this particular feat of genetic engineering is that such genes pass down the maternal line and do not become incorporated into pollen (which is male); they cannot therefore escape into wild relatives by cross-pollination. Genes for drought and salt resistance have already been successfully transferred into the chloroplast genome. This kind of genetic modification not only avoids any possibility of cross-pollination but also promises other substantial benefits. The need for antibiotic marker genes is eliminated because the inserted genes are already more precisely located in the cell, and much larger concentrations of protein are manufactured in each cell. These high yields make the technique an ideal method for developing edible vaccines and for the production of other pharmaceuticals in plants, taking the prospect of turning plants into factories of pharmaceuticals an important step forward. The benefits of the new technique could in due course be substantial.

Sterile GM plants

Another way to prevent cross-pollination is to ensure that transgenic crops are sterile. That was the reason why so-called 'terminator seeds' were in the process of being developed. However the label 'terminator', like the phrase 'Frankenstein foods', was a brilliant journalistic invention by the opponents of genetic modification, with its associations of danger that scared the public. If such transgenic crops had been marketed (which they never were), farmers would have been compelled to buy new seeds each year, a prospect that had greater appeal for the manufacturing company than for farmers. In the aftermath of Monsanto's ill-advised advertising campaign in Europe, Gordon Conway, President of the Rockefeller Foundation, persuaded the company not to pursue the development of the technology, and invited the agricultural seed industry as a whole 'to disavow the use of terminator technology', which it promptly did. Like the allergenic Brazil nut protein engineered into a soya bean, a technology that was never developed has been used ever since to frighten the children.

Terminator seeds may in fact make a comeback in more friendly guise. The environmental organization, English Nature, has urged that we should incorporate 'genetic incompatibility' into crops, that is, return to the original aim of engineering seeds to be sterile. It would provide 'a bio-safety mechanism, rather than a means of brand protection'.[20] It could provide an option for farmers worried about cross-pollination from their crops. In fact it would be a new version of 'terminator' crops aiming to achieve the purpose for which they were originally designed.

Biodiversity and pest-resistant crops

Pesticide usage

It is claimed that GM technology will harm biodiversity because pest-resistant crops will kill non-targeted insects as well as targeted ones, that is, beneficial insects as well as pests.

The evidence to date—from the United States, China, and South Africa—suggests that biodiversity in fact gains from the cultivation of GM pest-resistant crops because pesticides need to be applied less often. Typically US farmers report that they do not have to spray any pesticides on fields of *Bt* maize.[21] The change to pest-resistant GM soya bean, canola (rapeseed), cotton, and maize is estimated to have reduced the use of pesticides by 22.3 million kg of formulated product.[22] Since the introduction of *Bt* cotton into China, farmers have used 78,000 tons less of formulated insecticide in 2001 than before; the reason why small-scale Zulu farmers report greatly increased profits from *Bt* cotton is that they no longer have to buy expensive pesticides.[23] Furthermore, the Cotton Research and Development Corporation of Australia in its annual report of 2002 said that broad spectrum pesticide use on cotton farms had fallen 70 per cent since the first GM crop had been introduced seven seasons earlier.[24] In Mexico, where *Bt* cotton was first introduced in 1997 and now constitutes one-third of the total cotton production, insecticide use has declined by 80 per cent.[25] There is therefore a wealth of evidence about the beneficial environmental effects of pest-resistant GM crops.

Non-targeted insects

Do pest-resistant crops kill beneficial insects as well as pests? The most widely publicized allegation of harm to particular non-target insects was the story that *Bt* corn caused harm to the Monarch butterfly (*Danaus plexippus*). This butterfly has always exercised a special fascination for nature lovers; its long migration of thousands of miles, its ability to find its way back to the place it started from, and its beauty have made it a symbol of the complexity and wonder of nature. Alleged harm caused to the butterfly became a central feature of the campaign against genetic modification. Thus, a fundraising letter from The Environmental Defense Fund in the United States opened with the following remarks:

Each year millions of black and orange Monarch butterflies migrate thousands of miles from Mexico to Canada ... but new scientific evidence suggests that the milkweed Monarch caterpillars are eating may kill them and unless we act now, there could be more genetic surprises in the future.

A postcard issued by the Sierra Club featured a collage of Monarch butterflies with the words: 'Genetically engineered food which even butterflies find it hard to swallow'.[26] Much play was also made by environmentalists in the UK of the danger to the butterflies.

With regard to *Bt* technology, the US Environmental Protection Agency concluded in 1995: 'the Agency can foresee no unreasonable adverse effects to humans, non-target organisms, or to the environment'.[27] However, in 1999, it was reported that heavy sprinkling of pollen from *Bt* corn onto milkweed leaves in a laboratory experiment killed the larvae of the Monarch butterfly that ate the leaves and it was argued that the data could have 'potentially profound implications for the conservation of Monarch butterflies'.[28] The study was highly controversial. The journal *Science* had refused to publish the article because reviewers cited methodological problems, such as lack of proper controls and lack of field data, and in particular that no details were given of the actual dose of *Bt* toxin used. The journal *Nature* then accepted it as 'scientific correspondence' and because of its value as hot news, tipped off science writers in advance of publication. As expected, the story captured the public imagination. The technology had, it seemed, been caught red-handed: a GM crop was killing one of nature's most beautiful creatures.

Soon after publication, most plant biologists at a meeting in Rome rejected the validity of the study. One scientist said that if he had used such methods, he 'would expect to be chopped into little pieces during peer review'.[29] Others described it as a worst case scenario, 'just as an airline crash is the worst case scenario for flying'.[30] A variety of field studies, as opposed to laboratory studies, commissioned in response to the report in *Nature* concluded that the impact of pollen from *Bt* corn on Monarch butterflies in the field, as opposed to the artificial conditions of the laboratory, was negligible and not substantially different from the effect of conventional corn. In laboratory trials with *Bt* pollen, densities even five times greater than typically found in cornfields also had no significant effects on the growth and survival of the larvae. In fact, in 1999, a year when 30 per cent of all corn grown in the United States was *Bt* corn, the population of Monarch butterflies increased by 30 per cent.[31] It should be noted that, despite these later studies, campaigners against GM crops, true to form, never cease alleging that GM crops kill Monarch butterflies. One commentator has written that, 'like miners' canaries', the adverse effect of GM maize on wild Monarch butterflies is a warning sign of the danger of the technology.[32]

Another report claimed that the breeding capacity and viability of another beneficial, non-target insect, the green lacewing (*Chrysopa carnea*), was affected by *Bt* toxin, but the amount of toxin administered to the lacewing's prey in the laboratory was ten times higher than that found in any *Bt* corn in the field. This too was a worst case scenario. As emphasized earlier, *Bt* toxic proteins are extremely specific in their action and those in *Bt* corn are selected to kill moth larvae, such as those of the European corn borer, not the larvae of lacewings or ladybirds, or spiders or parasitic wasps—or honey bees. Populations of the predators of the pest would of course diminish as their host populations decline. In fact, when lacewings were given a choice, they showed an 'almost unanimous disregard' for the caterpillars that were dying because they had been fed on plants containing the gene for *Bt* toxin.[33]

It is perhaps surprising that anti-GM campaigners have concentrated on the *Bt* toxin, a substance that organic farmers are happy to spray on their own crops in unmodified form. The insertion of the gene for *Bt* toxin into crops should in fact be welcomed by environmentalists, because it is a more selective way of dealing with pests than the alternatives. As a distinguished molecular biologist has pointed out: '*Bt* insecticidal proteins selectively kill some beetles and caterpillars and target insects that eat crops. Expression of *Bt* protein into cotton and corn has reduced the application of specific, highly toxic pesticides by more than 80 per cent, allowing a substantive return of wildlife to crop fields.' Is this not, he asks, a giant stride towards Rachel Carson's goal of eliminating pesticides? Who are the environmentalists now?[34]

Resistance to Bt protein

ActionAid and other critics have argued that any success of *Bt* crops is bound to be short-lived because insects targeted by the various *Bt* toxic proteins will develop resistance. Contrary to expectations (largely due to management strategies to prevent the development of insect resistance), a study funded by the US Department of Agriculture found that no such resistance has yet been developed in the seven years or more that *Bt* plants have been grown commercially.[35] Furthermore, the development of resistance can be guarded against by growing refuges of unmodified plants beside *Bt*-containing crops. Any insect pests that survive the toxin are more likely to find a mate among the much larger populations of their non-resistant kind that infest the neighbouring plants which have not been exposed to the toxin, so that their offspring are unlikely to develop resistance. It was feared that many farmers, especially in poor countries, would ignore the need for refuges. However, for whatever reason, the continued effectiveness of *Bt* toxic proteins in the transgenic crops exceeds all expectations.

The effect of *Bt* plants on the environment has been the subject of a judgment by the federal court in the United States. Greenpeace International and a number of other campaigning organizations filed a lawsuit against the US Environmental Protection Agency (EPA) claiming that *Bt* plants cause unreasonable adverse effects, namely the evolution of *Bt*-resistant insects, gene flow to weedy relatives, and harm to non-target organisms. In reply, the agency stated that, while the Diamondback moth had developed field resistance from excessive foliar spraying and insects had been made resistant in the laboratory, it found no confirmed evidence that insect predators of corn, potato, or cotton have developed field resistance to *Bt* since the 1995 registrations of *Bt* crops. It found no significant risk of gene capture and expression by weedy relatives of corn, potato, or cotton. It knew of no data to support ecological risk to beneficial insects or soil. A federal court dismissed the lawsuit in July 2000.[36]

Biodiversity and herbicide-tolerant crops

Benefits from the development of such crops have already been listed, no-till agriculture probably being the most outstanding. Their cultivation generally requires less spraying of a range of herbicides than unmodified crops, allowing weeds to grow for longer before they are destroyed and allowing more insects to live on them. This was clear in Britain, for example, from an experiment at Broom's Barn in Suffolk, where herbicide-tolerant sugar beet was farmed in a way that substantially increased the insect and wildlife population.[37] However, recent field-scale trials conducted on a large number of farms in the UK between 2001 and 2003 gave results that caused some confusion, perhaps because of the way they were reported. The reports suggested that, compared with conventionally-farmed oilseed rape and sugar beet, their herbicide-tolerant equivalents supported fewer butterflies, and also fewer bees in the case of the rape. By contrast, more butterflies and bees were found in fields of herbicide-tolerant maize than in the controls.[38] Here, it seemed, was evidence from a carefully-conducted experiment which seemed to show that herbicide-tolerant crops could damage biodiversity.

The co-ordinator supervising the trials pointed out that what this proved was not that the GM crops themselves caused harm to wildlife, but that the herbicidal management of two particular kinds of GM crop grown under particular conditions had resulted in greater weed reduction and, as a consequence, in a reduction of the number of insects, including bees and butterflies. While this had been the effect when the crops were managed as they were in the trials on the farms concerned, this did not prove that other ways of farm management would not have a different result. Using the methods of planting in the Broom's Barn experiment, for instance, or providing wider margins at the side of fields where weeds could grow undisturbed, or growing more hedges, would all allow more wildlife to survive. It should also be pointed out that the differences between crops in the trials were bigger than the differences between GM and conventional crops. Furthermore, the Government in its wisdom has now apparently enshrined the principle that killing weeds among crops is equivalent to environmental harm.

The results of the trials were greeted by the Green lobbies as a vindication of their stand against GM crops: Friends of the Earth said they confirmed that GM crops harmed the environment, and Greenpeace immediately called for a total ban on their cultivation. Once again, inaccurate generalization based on dogma was preferred to objective analysis of evidence.

'GM crops are irrelevant to the needs of the developing world'

There are certain common themes in attacks on GM technology, of which the report 'GM crops—Going against the Grain' is again a typical example. They invariably include the statement that the real causes of hunger and poverty and disease have nothing to do with shortage of food or lack of modern technology. First, it is claimed that there is already enough food in the world to feed the hungry; this ignores the problems of distribution and the damaging effect on local farming. Nor does it acknowledge that we will have to feed another 3 billion people without the availability of extra land. Next, it is said that what poor farmers need is land, resources like water and electricity, affordable credit, rural extension services, access to local markets, decent roads, grain stores, and better infrastructure. The obvious answer is: yes, of course, who could disagree? But why not introduce new technology that can bring immediate benefit even before all these necessary changes have been made? It was fortunate that such arguments did not prevail before the Green Revolution was launched. However, the achievements of the Green Revolution are rather casually dismissed because not everyone benefited—millions of poor farmers were left out because they could not afford fertilizers and irrigation—and its gains 'were eventually offset by resulting soil erosion and the evolution of new diseases and pests'.[39] These factors have indeed lessened its effectiveness in recent years, but to conclude that its gains (reducing the numbers of the malnourished from a half to one-fifth of the world population at a time of huge population increase) were ultimately negligible is a strange comment from an aid organization.

ActionAid is equally cavalier in dismissing the well-attested success of Bt cotton in South Africa and China. Its report starts with a grudging admission that some South African small farmers have indeed obtained larger yields of cotton and have saved money on insecticides, but then alleges that small farmers do not understand the seed contracts, misunderstood the technology, and thought there would be no more need to spray. It further states that the farmers were not aware of the need to plant refuges and could easily fall into debt. The report sums up the undoubted success story of Bt cotton in South Africa by quoting another Green organization, GRAIN: 'Bt cotton may provide a small amount of relief to small farmers in the near term, but it threatens to make things worse in the end'. (It is noteworthy that Green lobbies always adopt the most pessimistic scenario and forecast doom—except when it comes to prospective food surpluses.) As for Bt cotton in China, this 'is killing the natural parasitic enemies of the cotton bollworm and increasing the number of other pests'.[40] So much for the success of Bt cotton in China.

Green lobbies accuse supporters of genetic modification of claiming that it can solve all the problems of the Third World. The case in favour may have been overstated, famously in Monsanto's oft-quoted advertisement of 1998 (see Chapter 4, p. 81). I know of no reputable plant biologist who makes such a claim. While the scientists see the potential of biotechnology, they are cautious in their forecasts. Few deny that many other technologies also have important parts to play, including former methods of farming whose lessons have been forgotten and forms of intermediate technology that some countries have never applied.

There is another myth that needs correction. 'GM Crops—going against the grain' and 'Feeding or Fooling the World' both imply that all Third-World representatives oppose the cultivation of GM crops. Yet what was notable about the OECD conference in Edinburgh in 2000 was the almost universal enthusiasm for GM technology shown by scientists from the developing world. Today South Africa, Argentina, China, and India give it strong support. Brazil has just legalized commercial plantings. Indeed, despite the relentless campaign against it by many NGOs, every year more and more governments in Africa are turning to biotechnology to help them solve their problems. Fortunately, other NGOs concerned with aid to the developing world—FARM Africa, for example—do not share ActionAid's dogmatic hostility to genetic modification.

'Multinational companies are a malign influence'

Opposition to GM technology in Britain and Europe is insepar-able from widespread distrust of multinational companies and their perceived role in its promotion. Opinion polls in Britain regularly show that consumers feel that they are the ones who take the risks and only big companies (and rich farmers) benefit. It is also a central theme of aid agencies and Green lobbies that multi-national companies are indifferent to the plight of small Third-World farmers and are incapable of delivering the crops that are needed because they are solely motivated by greed.

I do not share the obsession of some NGOs with the wickedness of the profit motive nor their assumption that it can never be combined with public interest, nor do I believe that multinational companies, whatever their occasional misdeeds, are inherently iniquitous (this subject is discussed in Chapter 9). Nevertheless it is, in my view, unfortunate that GM technology depends so extensively on development by corporations and that in recent years in most Western countries public finance for research into the plant sciences has steadily declined, and is still declining.

The new plants that made the Green Revolution possible and saved millions of lives were developed specially for the Third World by public institutions and by research financed from public funds, and most new crops that will bring the greatest benefits to developing countries today are still financed and being developed by public institutions. The dominant role now played by corporate research, or research dependent on corporate finance, has inevitably led to a concentration on products for the markets of rich countries. That is why the GM crops in widest use today are herbicide-tolerant crops and why three-quarters of the acreage on which GM crops are grown are in the rich industrial countries. Public attitudes in Europe might have been less hostile towards genetic modification if the first GM crops had been pest-resistant crops that benefit poor countries as well as rich ones (or if they had been products of direct benefit to consumers).

However, the corporate contribution to biotechnology in the developing world should not be ignored. Multinational companies have made the licences to develop Golden Rice freely available, as well as information about the sequencing of the rice genome. It was Monsanto and the Rockefeller Foundation which together launched the biotechnology industry in China that has since developed on its own. *Bt* cotton, which benefits China, South Africa, and India today, was first developed by multinational companies. Joint initiatives have also been taken by Monsanto and Rockefeller to encourage biotechnology in East Africa. Perhaps what is most important is that as developing countries become wealthier, their markets will begin to matter more. Hunger and disease both cause poverty and are caused by it. Enlightened self-interest, quite apart from any motive of social responsibility, will increasingly drive multinational companies to develop crops that help eradicate poverty, hunger, and disease. Indeed, the next generation of GM crops, such as plant vaccines orally administered and rice modified to contain a gene for iron, are crops directly relevant to the needs of the developing world.

The nature of the opposition

Of the many scientific innovations that arouse suspicion at the present time, none raises more passion than genetic modification of plants. It is understandable that a new technology of this kind should not be accepted uncritically. It is reasonable that questions should be raised about its safety, that doubts should be raised about its effect on the environment and that evidence should be demanded that it benefits rather than harms developing countries. But Green lobbies have abandoned reason as the basis for their opposition. They react to GM crops much as the religious right in the United States reacts to abortion. Their fanatical determination to stop its development blinds them to the moral consequences of their actions. Some of them will use any argument that comes to hand or raise any scare that may serve their purpose, rather like the eighteenth-century campaign against Jenner's smallpox vaccine, when people who received the vaccine were warned that cows' heads would grow out of their arms. (This drastic outcome is portrayed in a famous James Gillray cartoon, 'The Cow Pock, or the Wonderful Effect of the New Inoculation' of 1802, which lampoons the claims made by Jenner's adversaries.)

Two examples illustrate the nature of the opposition at its most extreme. One is the opposition to the project, mentioned earlier, to protect Bolivian potatoes against nematodes. The project itself is proof against all standard objections to GM crops. No multi-national companies are involved: the research has been publicly funded at Leeds University in the UK. No international politics are involved: the Bolivian government supports the project. Bolivian farmers will not be dependent on seeds sold by a big company exercising monopoly powers; there is no threat of monoculture; the benefits will go exclusively to small farmers and those who buy their produce. The diet, health, and incomes of Bolivians will benefit substantially. There can be no conceivable danger to health as the gene inserted into the potato is already found in the saliva of human beings. Patent rights do not come into the picture. The environment benefits, because new crops will grow on land exclusively used for potatoes today. No wilderness is threatened.

Nevertheless, activists in Bolivia try to stop local farmers planting the GM potatoes and use every means, however unscrupulous, to exploit the fears of unsophisticated people. Farmers have been told that eating transgenic potatoes may cause potatoes to grow from their heads or may cause unwanted pregnancies.[41] History repeats itself.

The second example is the campaign waged by Greenpeace against Golden Rice. This has been exposed by Ingo Potrykus, who has devoted more than a decade to trying to solve an urgent and previously intractable nutritional deficiency that affects the poor of the world. With his colleague, Peter Beyer, he has modified rice so that it contains enough pro-vitamin A to prevent vitamin A deficiency. I have already described the potential importance of this achievement. With the help of Zeneca, they have managed to overcome the formidable obstacle of dealing with no less than 70 intellectual and technical property rights belonging to 32 different companies and universities. The biggest remaining obstacle is Greenpeace and its allies.

As in the case of the Bolivian potato project, none of the standard Green objections to GMO crops can possibly apply. Potrykus has pointed out:

Golden rice has not been developed for or by industry. It ... [complements] traditional interventions. It presents a sustainable, cost-free solution, not requiring other resources. It avoids the unfortunate negative side effects of the Green Revolution. Industry does not benefit from it. Those who benefit are the poor and disadvantaged. It is given free of charge and restrictions to subsistence farmers ... It does not create advantages for rich landowners. It can be re-sown every year from the same harvest. It does not reduce agricultural ... or natural biodiversity. There is, so far, no conceptual negative effect on the environment ... etc.

Yet it is still passionately opposed by Greenpeace, which misrepresents its effectiveness (see Chapter 4, p. 85) and argues that it is not needed, that Vitamin A deficiency can be cured by other means, and so on. Some of the opponents of Golden Rice even argue that nobody wants it because it tastes awful or that people who eat Golden Rice will lose their hair and their sexual potency.[42]

At heart the opposition to GM is not rational but political, dogmatic, and ideological. It represents the triumph of eco-fundamentalism over reason.

6

The Rise of Eco-fundamentalism

Question:
'Your opposition to the release of GMOs, that is an absolute
and definite opposition . . . not one that is dependent on fur-
ther scientific research?'
Lord Melchett (Director of Greenpeace):
'It is a permanent and definite and complete opposition.'

Convictions are greater enemies of truth than lies.

Friedrich Nietzsche

The most common of all follies is to believe in the palpably
untrue.

H. L. Mencken

THE term 'fundamentalist' is one that is usually applied to the
adherents of a religion who base their beliefs and actions on the
literal interpretation of sacred texts. Thus, Islamic fundamentalists
base their beliefs and actions on the Koran; Jewish fundamentalists
justify their occupation of Palestinian lands by the words of the
Old Testament, and Christian fundamentalists (creationists) argue
that Darwin was wrong because his theory of evolution contradicts
the story of Genesis, the word of God. In that sense there are no
eco-fundamentalists, because there are no sacred ecological
texts—even *The silent spring* by Rachel Carson has not achieved
such status. On the other hand, some opponents of genetic modification have become so passionate in their opposition to GM crops
that they cannot be influenced by evidence and their minds are
firmly closed to rational argument. Since their beliefs have the
characteristics of a religion and their actions have much in com-

mon with an evangelical crusade, they can legitimately be described as eco-fundamentalists. Their form of environmentalism has come to be one of the most influential religions in the industrial world today.

Eco-fundamentalists must be distinguished from pragmatic environmentalists. The former often talk about the relationship of mankind and nature in mystical terms and show contempt for the evidence-based approach. They have become true believers, who regard science and technology as the enemy that threatens to destroy the environment. Pragmatic environmentalists are also concerned about the environment, but their concern takes into account the evidence of what is happening and they look at practical ways in which damage or threats to the environment can be repaired or avoided. They see science and technology as allies, not enemies.

Some organizations operate in a No-Man's-Land between the two groups. For example, I do not place aid agencies such as ActionAid in the fundamentalist camp, despite their biased and selective treatment of evidence, because I believe that in the end they are still open to persuasion. They are like advocates who are so convinced of the justice of their case that they believe that any evidence that contradicts it must be wrong. Unfortunately, while the aid agencies may not be fundamentalist themselves, some are clearly influenced by those who are.

Pragmatic environmentalism

Pragmatic environmentalists care about the environment partly from self-interest, partly for ethical reasons. For our own good, we cannot be indifferent to the environmental effect of our actions. If, for instance, we do not husband our natural resources, some may run out. It makes sense to place limits on the amount of fish that fishermen can catch in the North Sea and the North Atlantic because, if we do not, there will soon be no fish in those fishing grounds. If we do not care about pollution, our quality of life will deteriorate. Cities that expand higgledy-piggledy without any concern for environmental impact can soon become impossible to live or work in. We need to curb the expansion in the use of cars in cities to prevent traffic coming to a standstill. If we do not control air pollution, or sewage, or waste disposal, our health will suffer.

However, there are reasons other than self-interest as to why we should care for the environment. Love of beauty is one of them, which includes love of the natural world as well as love of art. A world without trees and flowers, birds and wild animals, clean air, rivers and seas, would be a world hardly worth living in. To be 'at one with' the natural world is a common human instinct. That is why, given a choice, so many people would like to live in the country. They feel at ease with themselves and the world when they are in surroundings of natural beauty.

Biodiversity matters too, because the infinite variety of species, both flora and fauna, is part of the glory of nature. This does not require us to take a static view of the natural world or regard the extinction of any one species as a tragedy: evolution occurs because species die out and new ones develop. Indeed, the total number of animal species in the world is unknown and can only be estimated. While there are over 20,000 known species of Chalcid wasps, the estimated number is some half a million; the number of species of beetles is estimated at two to four million. If a particular species of parasitic wasp or beetle dies out, some naturalists may mourn their passing, but life goes on. On the other hand, if it is true, as some ecologists claim, that we are losing some 40,000 species a year and that the impact of our present patterns of consumption and use of resources will lead to the sixth mass extinction in the history of life on earth, this should make us re-examine the way we live now. On a somewhat smaller scale, if by changing the way we manage our farms we can prevent the disappearance of skylarks, lapwings, reed bunting, or other species of birds from our fields, we should be ready to adapt our ways of farming. We have a moral responsibility to our children, our grandchildren, and future generations to hand over a world they too can enjoy.

Concern for the welfare of non-human creatures is also a moral responsibility. Those who are cruel to animals are likely to be callous towards their fellow human beings. Pain and suffering are not exclusive to the human species and one of the distinguishing features of mankind is that, unlike other species, we feel that our responsibilities are not confined to our own kind. Both aesthetic and moral considerations are therefore good reasons why we should care about the world around us and the creatures that live in it.

Animal rights extremists

One group of extremists who can be regarded as fundamentalist, in that they cannot be influenced by evidence or argument, are the extreme animal rightists. Animal rights campaigners are not satisfied with extensive laws that protect animals from cruel treatment. Antivivisectionists may argue (against the evidence) that performing experiments on animals for medical purposes is worthless and unnecessary, but their basic credo is that evidence of potential benefit to human beings is irrelevant. To them, all experiments using animals are immoral and can never be justified, even if they lead to the saving of human lives. An extreme minority go beyond claiming that animals have the same rights as us and believe that animal rights should take precedence. They do not shrink from using violence against those who perform, or are in any way associated with, animal experiments. They beat people up, terrorize their children, daub their houses with paint, firebomb property and set fire to cars, and resort to widespread intimidation. They regard any means that can stop research using animals as justified. This is not the first time in history that compassion for animals has been combined with indifference to human life. The Nazis passed antivivisectionist laws and, under the inspiration of Heinrich Himmler, the head of the Gestapo, the SS received special training in respect for animal life of 'near Buddhist proportions'.[1]

Such extremists have already succeeded in closing down several firms that breed laboratory animals for experimental purposes, whose business was not only lawful but necessary for medical research, and one group achieved national notoriety with a campaign started in the late 1990s to close down one particular company, Huntingdon Life Sciences (HLS). Its managing director was beaten up with a baseball bat; its employees were exposed to intense and often life-threatening intimidation. While this failed to deter HLS staff from continuing to work, the appearance of protesting placards outside banks and investment and insurance companies that provided financial backing to HLS soon persuaded these organizations to withdraw support. Pusillanimity in City boardrooms stands out in stark contrast to the fortitude shown by the HLS staff, and has encouraged the terrorists to believe that they are winning. Suppliers and clients of HLS have also been targeted but most of them have resisted intimidation.

Sadly, it seems that violence often wins. In January 2004 Cambridge University decided not to build a new laboratory to conduct research into neurological diseases in which some of the research would have involved work with primates. The reason given was the rise in the projected costs of running the laboratory. In fact, the main item of extra cost would have been the policing of the laboratory to prevent violence from animal extremists. If the handful of animal extremists win further successes, even more violent and intense campaigns will follow, directed against every University, company, or other organization that needs to use laboratory animals. Medical science would be gravely affected. Many people suffering from diseases that are now incurable would be deprived of knowledge or drugs that might help them. Biotechnology companies would be driven out of the UK. Our science base in biology, one of our national strengths, would be destroyed.

While animal rightists resort to terror and break the law, most antivivisectionists are idealists who disapprove of violence and law-breaking. Nevertheless their aim is the same, to stop the use of any animals in scientific procedures. If they succeed, they would do almost as much social harm as the extremists. They too would hold back the development of new life-saving drugs and undermine the study of one of the most important branches of science. Yet antivivisectionists receive support in some surprising quarters. Several investment institutions, which proclaim that they practise a policy of ethical investment, place companies that carry out experiments using animals, and indeed biotechnology companies as such, in the category of 'unethical' investment. They seem to be unaware of the needs of medical research, of the care that scientists show for the animals they work with, or the detailed control of animal experimentation exercised by the Home Office Inspectorate in the UK.

Animal rights vs. human rights

Leaving aside religious or semi-religious beliefs, are there any rational grounds for arguing that the rights of animals should be equated with the rights of human beings? Professor Peter Singer of Yale University, for example, argues that we infringe the rights of animals by using them for our purposes, and that one day this will be considered as immoral as slavery. In his view, an advanced primate, such as a chimpanzee, should be entitled to the same protection of the law as a human baby.

There are two main objections to equating animal rights with human rights. First, the view is based on a misconception of what rights are: what are frequently invoked as rights are really aspirations, like the right to work, the right to health, or the right to freedom from hunger. We would all like to see no one unemployed or unhealthy or hungry. But how can such 'rights' be enforced? What happens if governments ignore them, or find that circumstances prevent these aspirations being realized? A right that, in practice, no one has to respect is not a right, but an ideal. Meaningful rights are those that can be enforced. The European Convention on Human Rights, for example, now has legal effect in the UK and via the Human Rights Act 1998 imposes obligations on public authorities that can be enforced through the courts. If they are to be more than idle aspirations, rights must be anchored in obligations.[2] While we have legal obligations to protect animal welfare—those who are cruel to animals can be prosecuted—in general, talk of animal 'rights' is an example of loose thinking. Anyway, which animals have rights? Insects? Mammals perhaps? Does that include rats if there is a plague of rats? The more closely the concept is subjected to critical scrutiny, the more vague and meaningless or eccentric a comparison between animal and human rights becomes.

The second objection is based on a value judgment, one that most people would share: that our obligations to our fellow human beings are simply of a different order from our obligation to animals. This view is disputed not only by Western antivivisectionists: Buddhists and the followers of Jain believe that we have no right to extinguish any life, animal or human, including the life of insects, and Jain monks may wear muslin over their mouths to prevent them from swallowing flies. Ecologists who object to the anthropocentric view and take a rather long-term perspective of life on earth regard the possible extinction of mankind as no more tragic than that of any other species. Indeed, some who call themselves eco-centrists talk of mankind in positively derogatory terms, as a cancer on earth, because we are endangering the whole future of the planet.

If I have to choose between the death of a malaria victim and that of the mosquito that transmits the malarial parasite, I have no qualms about choosing death for the mosquito. Of course, a mosquito is a less appealing form of animal life than a mammal. A mammal can more obviously feel pain than an insect, if insects can feel pain at all. It may be argued that the example of the mosquito is unfair, since the choice between the interests of people and other mammals, or sometimes between the interests of people and the environment generally, can be much more difficult than choosing between a person and a mosquito.

What if poor Third-World farmers can only make a living by clearing forest land that is a habitat for rare wildlife, including mammals? An easy answer is that we must avoid that choice by helping them stave off poverty and starvation by other means. But suppose we fail to do so, as we have failed to do so far? In that case, I believe that people must come first: we have no right to condemn the farmers to poverty and hunger. I am more concerned with people than trees, or for that matter the fauna that live in trees. On the other hand, if whales are in danger of extinction, I would, like many others, place the need to preserve the whale before the interests of whalers because, among other considerations, the extinction of the whale would be a sad loss to the diversity of life around us. The cruelty involved in whale-hunting is another reason for banning it. Ivory-hunters, many of whom are poor, benefit from slaughtering elephants for their tusks, but most people support a ban on the trade in ivory. To favour a ban on whaling or on the trade in ivory does not, however, mean adopting the eco-centric view. It means taking account, as human beings, of our interest in the preservation of nature and in civilized behaviour that outlaws cruelty. As often happens when different ethical values conflict, we have to strike a balance, in this case between our love of nature, including other animals, and our concern for people, but we still make the judgment as human beings and from the human being's point of view.

The credo of eco-fundamentalists

Eco-fundamentalists make one exception to their general suspicion of science: they claim to base their world view on ecology, that is the study of organisms in their own environment. Ecology, since its foundation by Ernst Haeckel in the latter half of the nineteenth century, has always had a strong evangelical streak and Haeckel, who rejected the anthropocentric view of the world, was seen by many as the father of a new religion. Many ecologists have expressed a belief in the mystical unity of Mankind and Nature and have argued that the birth of science brought a mechanistic, rapacious, and inorganic attitude towards nature. Ecologists themselves are the 'saved', the only ones who know that Man (indeed very much Man as distinguished from Woman) is responsible for the impending apocalypse.[3] They hanker after a lost Arcadia, when people lived with nature in balanced eco-harmony, taking from Mother Earth only what they needed. But science and the Enlightenment destroyed Arcadia. To quote a more contemporary view,

between the 16th and 17th centuries, the image of an organic cosmos with a living female Earth at its center gave way to a mechanistic world view in which nature was to be reconstructed as dead and passive, to be controlled and dominated by humans.[4]

As the writer Michael Crichton has pointed out, the new religion is an almost perfect re-mapping of traditional Judaeo-Christian beliefs and myths: it has its own Eden and paradise, when mankind lived in a state of grace and unity with nature; then came the fall after eating from the tree of knowledge (science), and as a result of our actions there is a judgment day coming for us all in this polluted world. But true environmentalists will be saved, by achieving sustainability. To complete the analogy, he describes organic food as the pesticide-free wafer which the faithful ingest as part of their ritual of communion.[5]

Apart from rejecting anthropocentricity, the credo of eco-fundamentalists has three central tenets: (a) a belief in the unity of mankind and nature, which has been destroyed by science (at least narrow 'reductionist', as opposed to 'holistic', science) and by the handmaiden of science, technology; (b) a certainty that we are heading for ecological meltdown; and (c) a cavalier attitude to scientific evidence. A good illustration of these beliefs can be found in a book entitled *Rising tides* by Rory Spowers,[6] described in *The Daily Telegraph* as the Green equivalent to Naomi Klein's anti-globalization polemic *No logo*. *Rising tides* is not a work of any distinction, but I refer to it because it is typical of the eco-fundamentalist thinking to be found in periodicals such as *The Ecologist* and in much of the literature put out by organizations such as Greenpeace and the Soil Association. I believe most of the views expressed in the book accurately represent those held by eco-fundamentalists in general. They have gained widespread currency and considerable influence.

To start at the end, the book finishes with a call for an act of mystical faith that anyone but a dedicated believer in Deep Ecology and in the need to re-establish our Oneness with Nature might find somewhat difficult to achieve and which goes rather further than the normal injunctions of ecologists. To save the world, the author argues, we must evolve from ego-centric to eco-centric consciousness.

When the notion of a separate 'skin-encapsulated ego' is abandoned, replaced with a sense of 'ecological self'—which embraces not only our own bio-region but the universe itself—then we can make the quantum leap to an Ecological Age. ... When the conceptual being dissolves in the Ground of All Being, the conceptual separation between Man and Nature dissolves with it.[7]

Leaving aside such acts of true self-sacrifice, before he reaches this moment of redemption, the author, like many of his fellow-believers, regrets that science, in the form of modern agriculture, has turned its back on what was once a golden age of farming. After the Second World War, 'Thousands of years of traditional knowledge, encompassing methods for conserving nutrients in the soil and combating pests through diversification, were rapidly replaced by short-term chemical applications'.[8] There is a reference to the disasters caused by 'the chemical excesses of the so-called Green Revolution'[9] (an agricultural revolution which incidentally saved hundreds of millions from starvation). The past, pre-scientific age was always better. Spowers quotes Daniel Quinn: 'tribalism is not only the pre-eminently *human* social organization, it's also the only unequivocally *successful* social organization in human history'.[10] There are numerous references to the 'mechanistic' or 'mechanical' approach, which has taken over from the organic world view[11] and he cites Vandana Shiva, an Indian eco-feminist, a fanatical anti-biotechnology campaigner and one of the heroines of the fundamentalist cause frequently quoted by them as spokeswoman for the Third World, on the need to open up the intuitionist side of our being, which has been repressed by our preoccupation with reductionist science and our concentration on the analytical, rational side of our brain. She wrote:

[Intuition] is a way of knowing in which your relationship, your connection with other species, with the plant, with the soil, with a cow, with a sheep, is so intimate, so deep, that there is instant communication . . . it is the soundest base of knowing in the world.[12]

Teddy Goldsmith, founder of the magazine that is the voice of eco-fundamentalism, *The Ecologist*, is quoted by Spowers for observing sarcastically: 'God obviously did a bad job, and it is incumbent on our scientists to rearrange our universe according to their vastly superior design'.[13] Reductionist science, 'which has become so focussed on the building blocks of life that it is blinded to the mysteries of life itself', is once again the villain of the piece.

The trouble with intuition, reliance on traditional, pre-scientific knowledge, or the abandonment of 'reductionist' science in favour of the holistic approach, especially if it requires a spiritual dimension, is that there is no longer any objective test of truth. Why prefer one man or woman's intuition to another's? What evidence can prove or disprove the tenets of 'holistic' science? Why not embrace astrology? If you rely on intuition, how can one form of fundamentalism argue that it is rationally to be preferred to any other? This is one reason why different fundamentalist faiths regard each other with intolerance, if not with hatred, and often end up killing each other.

Another central theme that pervades the credo of *The Ecologist* and all eco-fundamentalists, and which again *Rising tides* accurately exemplifies, is the conviction of coming doom. Our attempt to control nature through science and technology is, they feel, endangering the future of the human race and facing all life on the planet with extinction. There are repeated references to ecological melt-down and continuing echoes of the predictions of 'Limits to Growth', that we are rapidly running out of natural resources. The world is in the situation of 'being in a huge car driving at a brick wall at 100 mph and most of the people are arguing where they want to sit'.[14] The entire mantra of the doomsters is recited as established fact: Rachel Carson was right about the carcinogenic effects of chemicals like DDT. 'We shall never know how many human lives have been cut short by exposure to these chemicals, but the list of cancers and modern diseases which can be attributed to them grows longer every year'.[15] (In fact there is no evidence that DDT causes cancer in people, whereas it cannot be disputed is that it saved tens of millions from death.) Again: 'recent studies suggest that up to 90 per cent of modern cancers can be attributed to dietary and environmental factors.' And it is reported as a fact that 'the vast increases in various cancers, as well as the surge of modern diseases like Alzheimer's, is being related to ... synthetic chemicals'.[16] So it goes.

Sometimes it seems that eco-fundamentalists would actually be deeply disappointed if they were proved wrong and science did manage to provide a solution to the problems of the world now leading us to doom. Global warming is nature's revenge for our wickedness in thinking we can control nature. We are being punished for our greed. Pollution is the modern pestilence that Isaiah blamed on our godless ways. Only if we adopt the faith that preaches ecological purity and sustainability can we be saved.

Attitudes to evidence

In his book *Voodoo science*, Robert Park listed seven warning signs that claims made for research results should be treated with profound suspicion (see Chapter 2, p. 57). A similar list of indicators should warn us that claims made by extreme environmentalists are likely to be spurious. One is the citation of statements or 'research' done by fellow-believers, without taking into account any contradictory evidence. The second is the making of general statements of such vagueness that their content cannot be measured and therefore proved or disproved. A third is to refer to 'recent studies', or make statements such as 'scientists say' or 'it is now generally agreed', without specifying what the studies or who the scientists are, or on what evidence such 'general agreement' is based.

All three indicators are liberally scattered throughout the pages of *Rising tides*. The most frequently quoted sources are leading, well-known environmental campaigners of the extreme kind, such as Vandana Shiva, Jeremy Rifkind, Rupert Sheldrake, and the magazines *The Ecologist* and *Resurgence*. At no stage is there any reference to any finding that contradicts claims of environmental damage or risk to health. No mention is made of any of a number of relevant reports from the Royal Society, or the Food Standards Agency, or the American Food and Drugs Agency, or the Centers for Disease Control, which show, for example, that there is no evidence that DDT is carcinogenic or that permitted levels of pesticide residues in food constitute any danger to health. There are no references to publications in *Nature* or any respectable scientific journal that requires papers to be peer-reviewed. There are numerous broad, unsubstantiated generalizations about diseases being caused by dietary and environmental factors. The text is littered with statements such as: 'Despite numerous studies which prove . . .', 'Studies show . . .' and 'Some studies suggest . . .', without any identification of the studies or of the scientists (if any) who conducted the research. The author even goes further and asserts dogmatically: 'It is *beyond doubt* that the global implementation of agricultural systems like organic farming could feed the expanding population [of the world] without the need for chemicals or biotechnology'[17] (my italics). Doubt does not feature prominently in the fundamentalist vocabulary. When evidence is quoted, it is strictly for show.

Green lobbies

The Green lobbies represent a somewhat mixed bag. Some are pragmatic environmentalists who seek to ensure that respect for the environment remains high on the government's agenda. However, many Green organizations have adopted an essentially fundamentalist approach. Greenpeace is a good example. In my Prologue, I traced the transition of Greenpeace from an environmentalist lobby campaigning for good causes that practical environmentalists could support to an organization whose desire for publicity and whose dogmatism has gradually become more extreme. Its rake's progress has been graphically described by one of its founders, Patrick Moore (not to be confused with the astronomer of the same name), who remains a committed environmentalist. As explained on his website,[18] when he was still a member of Greenpeace and it had first achieved sufficient standing to be taken seriously by governments, Moore argued for co-operation between government, industry, academia and the environmental movement. He found instead that his colleagues in Greenpeace preferred continued confrontation and ever-increasing extremism. He left the organization when it turned against science and technology and became, as he regards it, part of a hysterical left-wing campaign against globalization and free trade. The extent of his disillusionment is shown by his final comments in an interview with *New Scientist* in 1999:

Environmentalism has become codified to such an extent that if you disagree with a single word, you are apparently not an environmentalist. Rational discord is being discouraged. It has too many of the hallmarks of the Hitler Youth, or the religious right.[19]

The comparison with Hitler Youth is absurd, but the reference to the religious right is apt.

The fundamentalism of Greenpeace

The charge that Greenpeace, probably the most successful and influential Green lobby organization, has become to all intents and purposes eco-fundamentalist, should not be lightly made. Most of its supporters are not fundamentalists but subscribe to Greenpeace for commendable reasons, because, to oversimplify, they want to save the planet, or at any rate to stop its degradation, and believe the claims of Greenpeace that it campaigns on their behalf. Some of the causes it supports are indeed good causes. However, what most of its supporters may not realize is how ready Greenpeace is to play fast and loose with evidence in the pursuit of its aims, or indeed that it has become so convinced of its righteousness that it no longer regards evidence as relevant.

The fundamentalism of Greenpeace was illustrated by the answers given by its then director, Lord Melchett, to a House of Lords Select Committee on EU Regulation of Genetic Modification in Agriculture, which reported in 1999. He was asked:

Your opposition to the release of GMOs, that is an absolute and definite opposition? It is not one that is dependent on further scientific research or improved procedures being developed or any satisfaction you might get with regard to safety or otherwise in the future?' He answered: 'It is a permanent and definite and complete opposition based on a view that there will always be major uncertainties. It is the nature of the technology, indeed it is the nature of science that there will not be any absolute proof.[20]

The basis of this answer purports to be scientific, in that it refers to inevitable uncertainty. Lord Melchett was right to say that science can never provide absolute proof. It does not pretend it can. As David Hume pointed out, we cannot prove logically that the sun will rise tomorrow—although past experience makes it a reasonable assumption that it will. However, if the lack of absolute proof of the absence of potential harm were grounds for opposing new technology, we should oppose every technology, indeed dispense with the applications of science altogether. The important part of his statement was the admission (or perhaps it was a boast) that no evidence could ever change his mind or affect the opposition of Greenpeace to GM crops. Their position is therefore a matter of dogma or religious belief impervious to reasoned argument or evidence. It shows how hollow are the attempts that Greenpeace periodically mounts to 'prove' that GM crops are unsafe or damaging to the environment. Any evidence that contradicts its view can be ignored, however expert or independent the source, because it *cannot* be right, just as Darwinism cannot be right to creationists since it contradicts what is written in the Bible. Moore's comparison with the religious right stands up.

Indeed, like true religious zealots, the more enthusiastic members of Greenpeace want deeds, not words, and have stirred the movement into physical activity against the enemy. When farm-scale trials of GM crops were set up by the Government in 2000 to test their environmental effects, Greenpeace crusaders, clothed not in white armour but in white suits to protect themselves against assaults by hostile chemicals, were shown on television venturing forth into fields where GM crops were grown, led not by a mediæval knight but at least by a contemporary baron, Lord Melchett himself, to uproot the enemy out of the soil. Some raiders (not on this occasion Greenpeace) attacked the wrong enemy—a case of collateral damage from 'friendly fire'. In one battle GM trees were destroyed that had been developed to reduce the use of bleach and energy in the manufacture of paper, but the trees were female. Therefore they could not produce pollen and 'contaminate' other trees. They were the unfortunate casualties of war (and revealed the ignorance of the crusaders). Greenpeace was not just unconcerned with evidence or finding out the facts; in true mediæval style, it applied the logic of those who burnt witches, that you must destroy them before you determine whether they can actually cause harm.

Even less defensible than destroying farmers' property were the tactics some fanatics (not, I should add, identified as Greenpeace members) borrowed from extreme animal rightists. Several farmers who agreed to carry out field trials have been terrorized. People living in villages near proposed GM crop trials were told that if they let the crops be grown their children would suffer from cancer or allergies. Farmers have received hate mail and farmers' families, as well as their houses, have been threatened. Scientists who take a public stand in favour of genetic modification have received threatening letters, including a bomb threat.

Another example of its dogmatic attitude is the campaign by Greenpeace against the building of new state-of-the-art incinerators. These emit a tiny, insignificant quantity of dioxins into the air compared with the older incinerators which, particularly in the case of those burning hospital waste, caused concern to environmental agencies because of the quantity of harmful chemicals they released into the air. The new incinerators also avoid the need for the disposal of waste in landfill sites that often involves transport of waste by lorry over long distances. Many of them form part of new combined heat and power plants that are one of the most efficient means of generating low carbon energy. However, any new incinerators provide opportunities for television publicity, because intrepid Greenpeace mountaineers can be filmed scaling their high chimneys to display large banners from their tops.

Protesters invoke the spectre of a massive release of cancer-causing dioxins. Dioxins are never mentioned by eco-warriors, or indeed by those journalists who accept their propaganda, without being described as 'the deadliest chemicals ever made'. They are organochlorine compounds and Greenpeace has declared war on all organochlorines, calling chlorine 'the Devil's element'(note the religious overtones). Yet, to quote the prize-winning science writer and chemist John Emsley: 'no member of the general public has ever died of dioxin poisoning, despite the fact that for 40 years the chemical industry inadvertently produced large amounts of dioxins as impurities in other products'.[21] Emsley points out that there have been several major accidents resulting in the release of dioxins (e.g. the Seveso explosion in 1976), yet the number of workers who probably died from heavy exposure to TCDD, the most toxic form, was only four. Dioxins are naturally present in the environment. We eat them, breathe them, and drink them daily. He quotes Professor Christopher Rappe, the world's leading researcher into dioxins: 'More people make their living from dioxins than suffer from them'. Amongst the beneficiaries Professor Rappe includes lawyers and those who make very effective political capital from the compounds.[22]

In fact, there is increasing evidence that a small quantity of dioxins instead of doing us harm, may actually be good for us. Studies in toxicology show, perhaps paradoxically, that toxins that damage health at high concentrations may be beneficial at low concentrations, due to the 'hormesis effect' (see Chapter 3, pp. 72–3). The toxins include potential carcinogens like dioxin and cadmium, which, encountered at low doses, may actually reduce cancer rates.[23] The studies may cause us to review numerous safety regulations designed at some cost to eliminate pesticides or toxic chemicals altogether and would make nonsense of the Greenpeace campaign that 'even one dioxin is one too many'.

Confusion about sustainability

Ever since the publication of *Limits to growth* (and indeed before then) environmentalists have been concerned, at varying levels of intensity, about the conservation of limited resources and about the kind of world that future generations will inherit. Pragmatic environmentalists share the concerns about dwindling stocks of fish, threats to the survival of various birds and mammals, the possibility of serious harm to various parts of the world from climate change and a variety of potential threats to the quality of life that we now enjoy. Being pragmatic, they will react according to the best available evidence that the threats are real and adopt remedies that are appropriate. More pessimistic environmentalists, and especially the eco-fundamentalists, are convinced that only drastic changes to our way of life in the immediate future can avert doom. Salvation depends on achieving 'sustainability' and in the eyes of the fundamentalists those who question belief in our headlong rush towards ecological disaster are either blind or wicked. Arch villain is Bjorn Lomborg, author of *The skeptical environmentalist*, who argues that the world around us is not getting worse but generally getting better.[24] The debate about his book goes to the heart of the dispute between optimists and pessimists.

However, before analysing the reaction to Lomborg's book, it is worth asking what 'sustainability' means. The Brundtland report on *Our common future* in 1987, the Rio de Janeiro Earth Summit in 1992, and the Johannesburg Conference in 2002 were set up to promote 'sustainable' development, 'sustainable' production, and 'sustainable' consumption, and any number of national and international bodies are now in existence to promote 'sustainability'. But like the fashionable Precautionary Principle, the concept is more often invoked than defined. Many corporations, it seems, now have computer software programmes that automatically insert the word 'sustainable' before 'development' in every document.[25] In a recent edition of a magazine called *Green Futures*, the word 'sustainable' or 'sustainability' occurred no less than 71 times in six articles covering nine pages.[26] The term is scattered so liberally that it has virtually been deprived of meaning, except as a form of religious salvation.

A definition of sustainability that is frequently cited as a good working definition appears in the Brundtland Report: 'sustainable development is development that meets the needs of the present without compromising the ability of future generations to meet their own needs'.[27] It has a certain common sense appeal, which corresponds to the general feeling that we must look after the needs of our descendants and must not deprive them of the enjoyment of the environment that adds to the quality of our own lives. But on closer inspection the Brundtland definition is too vague to be a clear guide for making policy. What are the needs of the present generation? Clearly many of them are far from being met, since 800 million people do not have enough to eat. Perhaps we should think of meeting present needs before we worry too much about the needs of the future. Furthermore, what will be the needs of future generations? How can we tell fifty or even a hundred years ahead?

One thing we can be fairly sure of is that they will be much less wealthier than us. If world GDP grows at a modest 1.5 per cent per annum—which is lower than the average growth rate of the last fifty years and much less than recent growth rates of most developing countries—in a hundred years' time, average incomes will be over four times higher than they are today. Perhaps a wealthier society will decide, sensibly, that quality of life matters more than spending money and will prefer leisure to higher earnings. How do we know? Future patterns of consumption decades ahead are entirely unpredictable. However, whatever they are, it is not necessarily ethical to ask the present generation to make sacrifices for future ones who are likely to be much better off.

It will be argued that the issue is not income per head in the future, but what happens to the world's climate and its resources and the quality of the environment, of the air, the seas, the forests and wildlife. If these are deteriorating and the damage is likely to be irreversible, it is right to take action now. However, in most cases the difficulty is that views about appropriate action change over the years. We have entirely different information about the depletion of resources today from that which was available thirty years ago. In 1970, aluminium reserves were estimated to be 1170 million tons. By mid-1999, the forecasts were 34,000 million tons.[28] In 1932, the much-respected economist A. C. Pigou argued that it would be wrong to use up

for trivial purposes a natural product which is abundant now but which is likely to become scarce, and not readily available, even for important purposes, to future generations. This sort of waste is illustrated when enormous quantities of coal are employed in high-speed vessels in order to shorten in a small degree the time of a journey which is already short. We cut an hour off the time of our passage to New York at the cost of preventing, perhaps, one of our descendants from making the passage at all.[29]

Doomsters, in their predictions of future shortages and disasters, make racing tipsters look as reliable as the Admiralty Tide tables.

One thing we can say with confidence is that our knowledge will improve, as will the technology available to deal with future problems. Nor should we forget that in dealing with pollution, wealthier nations do so more effectively than poorer nations. The rich world has reduced the amount of local pollution of the air and of its rivers to an extent that few could foresee. It is likely that environmental problems will be easier to solve in the future if most countries are much richer than they are today.

Global warming

The most serious argument for early action to prevent irreversible damage is based on forecasts of global warming. Any discussion of sustainability cannot avoid taking a view, however tentative, on the issue that is central to the environmental debate. My views are very tentative, because the subject is immensely complicated, since 'climate is the product of so many variables—rising and falling carbon dioxide levels, the shift of continents, solar activity . . . that it is as difficult to comprehend the events of the past as to predict those of the future'.[30] Yet it seems, as has been said of cosmologists, that many climatologists are often wrong but never in doubt.

On balance, I accept the view of the scientists on the IPCC (the Intergovernmental Panel on Climate Change) that global warming is happening. It appears to be confirmed by the retreat of glaciers, the increase of ocean temperatures, and a gradual rise in sea levels and is accepted by most climate experts. Next, there seems to be convincing evidence that a significant proportion of global warming is caused by man-made increases in carbon dioxide levels rather than changes in solar activity. The third report of the IPCC in 2000 concluded that 'there is new and stronger evidence that most of the warming observed over the last 50 years is attributable to human activities.'

More controversial are projections about the *degree* of global warming likely to take place. The IPCC projects an increase in temperatures over the next century between 1.4 and 5.8 degrees Celsius, but assigns no probabilities to figures within that range. The figures depend crucially on increases in energy emissions and these in turn largely depend on projections of economic growth. They are not therefore scenarios based solely on the best available scientific evidence, but also on an economic input. This input appears open to question.

The Special Report on Emissions Scenarios (SRES) prepared for the IPCC has been challenged by two distinguished economists, Ian Castles, former head of the Australian Bureau of Statistics, and David Henderson, former head of the Department of Economics and Statistics at the OECD, who argue that the SRES has made two errors: firstly, it has used the wrong measure for comparing the economic output of different countries in the base year of 1990, namely that of market exchange rates (MER), instead of the more widely used measure of purchasing power parities (PPP).[31] The choice of measure is important, because the former, MER, gives developing countries a much lower share of world GDP than if the latter (PPP) is used, and therefore presents a picture of far greater differences in GDP per head between rich and poor countries. Secondly, the SRES assumes, in key scenarios, that this large gap in GDP per head between developing and developed countries will be very substantially reduced and its assumption of convergence is explicitly based on what is seen as equitable—not so much on argument and evidence as on the grounds that the world would be a better place if it occurred. To achieve this convergence would require an extraordinarily rapid growth rate on the part of the developing world, far exceeding growth rates that have been achieved historically. It would also significantly increase the growth of the world's GDP as a whole. This assumption of rapid growth, to close an initial gap which is greatly overstated, is reflected in higher projected emissions and correspondingly greater projected global warming.

The consequences of the SRES assumptions can be illustrated by considering their least alarming scenario, that which yields the lowest increase in projected emissions. In that scenario, it is assumed that by the year 2100 the gap in GDP per head between developed and developing countries will have narrowed to a ratio of 1.8 to one. On the MER basis this means that, given the projected rate of growth in the rest of the world, the developing countries would have to increase their GDP between 2000 and 2100 by over 39 times, while the GDP of the world as a whole would increase by 12 times. On the PPP basis, the respective figures would be 15 times for the increase in the GDP of developing countries, not 39, and 9 times for the increase in world GDP, not 12. That is, world GDP would increase by at least a quarter less than on the MER basis. Furthermore, even in their lowest growth rate scenario, the SRES still assumes that in the next thirty years developing countries would grow at a rate of GDP per head that is nearly twice as fast as the rate they have achieved over the past thirty years.[32]

Castles' and Henderson's conclusion, which I find persuasive, is that there is an upward bias in the lowest projections of total world emissions of greenhouse gases.[33] Another critic of the IPCC projections, Robert Ehrlich, Professor of Physics at George Mason University, USA, likewise argues that the realism of the projection of economic convergence between poor and rich is open to question, but observes that in a forum in which developing nations are strongly represented it would be politically incorrect to say so. He concludes that the amount of global warming is likely to be at the bottom of the IPCC's range.[34] It seems a reasonable conclusion that we should not assume global warming will happen very fast or will be as severe as pessimists project.

However, could the effect of this still be so serious that we should take drastic action now? Even moderate temperature increases, if they lead to a rise in sea levels, can bring devastation to those living in river deltas or low-lying islands like the Maldives.[35] In other areas modest increases of temperature would probably have beneficial effects. Winter warming would exceed the rise in summer temperatures and since more people die of cold than heat, more people would live longer. Moderate warming would be generally good for agriculture. It may or may not be good for arid regions of the world, according to the IPCC report; it talks confusingly both of the risk that extremes of weather may cause droughts in some mid-continental areas and also of an increased availability of water in some currently arid regions.

Assuming that coastal areas will suffer, there is clearly a strong case for special aid to countries like Bangladesh to enable them to take preventive measures against floods, as the Dutch have done over past centuries. But more generally the case for immediate drastic measures hardly seems to be made out. In a decade the picture may be much clearer, when in any event technology will be more advanced to cope with the problems we face.

The reaction to *The skeptical environmentalist*

The vagueness surrounding the concept of sustainability and the uncertainties about the rate of global warming suggest that it is unwise to be dogmatic about the future of the environment.

Yet even academic environmentalists do not seem immune to dogmatism and the influence of green fundamentalists has clearly spread. This was demonstrated by the reaction to the publication of Bjorn Lomborg's *The skeptical environmentalist*. Its main theme is an assault on what he calls 'the litany', namely the constant reiteration by a number of leading environmentalists (echoed uncritically by many commentators in the media) that we are facing an ever-deteriorating environment and even ecological meltdown. Some ecologists tell us that, as human beings are stripping the land of its green cover, polluting the air, and poisoning the seas, this together with population growth must inevitably lead to mass starvation. Lomborg challenges the view, expressed in the words of *Time* magazine, that 'everyone knows that the planet is in bad shape.' Instead, quoting some 3000 statistics, mostly from official international sources, he argues that the state of the environment, while not good, is in fact getting better.

Lomborg cites evidence that people live longer in all parts of the world and that life expectancy has doubled over the last hundred years. We have more to eat: in the last thirty years, the average calorie intake per head in the developing world has increased by 38 per cent and the number starving has declined from 35 per cent to 18 per cent. The absolute, as well as the relative, number of the undernourished has declined, despite huge population growth. People in the world have become better educated. Air pollution in the developed world has diminished dramatically and, although it has worsened in the developing world as a result of economic growth, there is every reason to expect that in time greater prosperity will reduce pollution, as it has in the developed world. Rivers and coastal waters have become cleaner. Lomborg challenges the conventional view about the prospect of a mass extinction of species, which he regards as unproven, and argues that there are grounds for preferring the lower forecasts of the pace of global warming to the more pessimistic ones. Most controversially, he argues that instead of following the recommendations of the Kyoto Protocol, which, at vast cost, would postpone the adverse effects of global warming by only seven years in the next hundred, we should concentrate resources on providing poorer nations with better health and education, and cleaner water and sanitation.

No book published in recent times has caused a greater furore in the environmentalist world. It has been condemned in almost hysterical terms. Lomborg was accused by *Scientific American* of wilful ignorance and destructive campaigning, lack of even a preliminary understanding of the science in question and of producing nothing more than a diatribe.[36] The editor criticized his presumption in challenging the views on climate change of investigators who have devoted their lives to the subject.[37] How dare he challenge the voice of authority! There were echoes of Calvin's denunciation of Copernicus: 'Who will venture to place the authority of Copernicus above that of the Holy Spirit?' It is clear that in many quarters the story of man-made global warming has become almost as unassailable as the bible story in the American bible belt.[38]

In the eyes of some of his critics, Lomborg is not just wrong but wicked: he is not just selective in his citation of evidence, he is not even entitled to join in the argument, as he is not a member of the environmentalist club. A review in *Nature*, most of which consisted of abuse rather than discussion of his thesis, accused him of arguing like those who deny the Holocaust, and referred readers to the website of an academic who was so proud of having thrown a pie in Lomborg's face that he posted a picture of the event.[39] Because Lomborg questioned the extent of global warming, he was denounced as a right-wing apologist for the Bush administration and American corporate interests. In Denmark, a Committee on Scientific Dishonesty found his book to be 'clearly contrary to the standards of good scientific practice' and accused him of 'systematic one-sidedness in the choice of data and line of argument', although it admitted that it had not considered it to be their task to determine whether Lomborg or his critics are right, and seems to have relied almost wholly on the denunciation of his work in *Scientific American*. This finding has since been overruled by another government body.

I confess that I start with a certain bias in Lomborg's favour. He is a cyclist who does not own a car. He used to support Greenpeace, until he became disillusioned. He is concerned with world poverty and seems to be neither a believer in the virtues of an unrestrained free market nor in the inherent wickedness of all big business. He argues that our top priority should be the reduction of poverty in the developing world. He is not a scientist, but a statistician who cites a wealth of figures from highly reputable sources, such as the FAO, the WHO, various UN committees, and other international bodies, which appear, *prima facie*, to support his case. (I was hoping that the reviews of his book would indicate whether his use of statistics was accurate and fair or selective.) Like him, I believe that we ought not to let environmental organizations, business lobbyists, or the media dictate priorities or monopolize discussion of environmental issues.

One of Lomborg's 'crimes' is his optimism. I too regret the loss of optimism about science that characterized the Enlightenment and its transformation into contemporary pessimism. I have always associated science with optimism, because there must be a sense of excitement about the process of discovery. If you also believe that science can help answer many of the problems that face us, you are more likely to be an optimist than a pessimist. I do not regard progress as inevitable, but it is not impossible, and I believe we can make the world a better place. Doomsters have managed to convey a spurious impression of intellectual depth and persuade public opinion that pessimism is profound and optimism is shallow. They are popular as media pundits because good news is not news, whereas warnings of catastrophe sell newspapers and attract television audiences. The current widespread anti-science mood has been strongly influenced by warnings that nowadays every minute particle of a chemical residue in our food or every extra molecule of a dioxin in the air claims a new cancer victim. It was therefore refreshing to read Lomborg's full-frontal assault on the prophets of doom and to learn that perhaps the world as we know it is not coming to an end, but is actually improving.

However, sympathy for Lomborg or antipathy towards the tone of attacks on him is irrelevant. The question is whether he is right, or whether the evidence he presents at least demands serious consideration. Are his critics devotees of a religion who are shocked because their faith has been attacked by a heretic? Or is their hostility justified because he has distorted evidence or because he is an environmental equivalent of Dr Duesberg, who denied that there was a link between HIV and AIDS? Or is his thesis basically right while parts of it are wrong?

The case against Lomborg

It is obviously difficult for an outsider and layman to evaluate his use of statistics when reviewers are so passionately divided: one side represents his book as 'one of the most important contributions to public policy in recent times',[40] while the other argues that his comments are so destructive and ill-founded that Cambridge University Press should never have published the book.[41] As the review in *Nature* revealed, a whole industry is now devoted to debunking the book, chapter by chapter.

Several more rational and restrained critics argue that, while in many respects the environment is improving, as Lomborg states, many improvements, for example in air quality in London, are largely due to regulation, such as the Clean Air Act of 1956, enacted as a result of pressure from environmental lobbies. They have made a plausible case that Lomborg understates the importance of regulation and somewhat exaggerates the importance of technical progress. Even if he does, this criticism, if anything, strengthens his central thesis that meltdown is not inevitable, since it suggests that a combination of regulation and technological progress can avert catastrophe.

The most reasoned attack on Lomborg has been against his views on global warming. The economist Michael Grubb accuses Lomborg of neglecting 'the literature of the past 40 years [which] demonstrates unequivocally that developments and dissemination of technology respond to economic incentives, such as those embodied in Kyoto's commitments'.[42] Another economist, Adair Turner, who agrees that Lomborg effectively demolishes the anti-modernist, anti-capitalist view of the world, suggests that there is a basic contradiction in his argument: on the one hand, he says that, in time, renewable energy technologies will solve the problem of global warming through the operation of the market; on the other hand, he cites statistics to show that there will be no shortage of energy and probably no shortage of fossil fuels for several centuries. But if fossil fuels are plentiful, their price will be low, there will be no incentive for the rapid development of renewable energy and there will continue to be a high-level emission of greenhouse gases.[43] Again, Turner argues that the more optimistic view of the rate of global warming which Lomborg adopts depends on the forecast that world population will peak at some 9 billion in 2040 and will then start to fall back to some 7 billion by 2100. This sounds a somewhat optimistic assumption. Indeed, Lomborg himself quotes the UN's central forecast that the population will be 9.3 billion in 2050 and will stabilize just short of 11 billion people in the year 2200. Turner concludes that Lomborg's optimism about climate change is implausible. (At a lecture I attended in 2003, Lomborg in fact took a rather moderate view of the potential from renewable energy, which seemed to suggest that Turner's criticisms are overstated.)

However, even if these criticisms of parts of Lomborg's thesis are justified (and I have given my own reasons, based on the economic arguments of Castles and Henderson, for supporting the less pessimistic view), it seems clear that others are distorted by a gut hostility to his central theme. He has answered the attack made on him in *Scientific American* point by point on his website, since he was refused space to answer in the journal.[44] To an outsider, his reply is impressive. It shows that actual examples of inaccuracy in his book are few (two of them he admits), while the inaccuracies of his accusers in misquoting him are many. He points out that he had attacked two of the four reviewers in *Scientific American* in his book as contributors to the litany of doom: one of them, Thomas Lovejoy, had predicted twenty years earlier that 15–20 per cent of all species would have died out by the year 2000, while a second, John Holdren, had long argued that the world was running out of resources. Holdren has since somewhat changed his theme, which is now 'that we are running out of environment—that is out of the capacity of air, water, soil and biota to absorb, without intolerable consequences for human well-being, the effects of energy extraction, transport, transformation and use.' It still seems not a million miles from the litany of doom that Lomborg, as most agree, has successfully demolished. The third of *Scientific American*'s critics, John Bongaarts, also seems to keep a foot in the doomsters' camp, quoting Paul and Anne Ehrlich that feeding the world's population will 'turn the earth into a giant human feedlot', at immense cost to the environment.

The 'mass extinction' of species

Another criticism of Lomborg is that he is wrong to dismiss claims that we are facing a massive extinction of species. This is a serious charge, although these claims have themselves been seriously questioned. It is commonly asserted, as if it were established fact for example, that we are facing the sixth mass extinction in the history of the world. It is therefore worth looking at the way figures of this magnitude come to be adopted. Matt Ridley, the author of a much admired book, *Genome*, in a (very pro-Lomborg) review of *The skeptical environmentalist* explains the origins of the commonly quoted figure of 40,000 species becoming extinct each year.[45]

The number was first used in 1979 by the British scientist Norman Myers. Yet what was the evidence for it? Here is what Myers actually said: 'Let us suppose that, as a consequence of this manhandling of the natural environments, the final one-quarter of this century witnesses the elimination of one million species, a far from unlikely prospect. This would work out, during the course of 25 years, at an average rate of 40,000 a year.

That's it. No data at all; just a circular assumption: if 40,000 species go extinct a year, then 40,000 species go extinct a year. *QED*.

Part of the problem is that 'mass extinction' is an emotive phrase that obscures its lack of precision and certainty. To start with, we do not know how many species of animals there are. Estimates vary from three to one hundred million, with a central range of some five to fifteen million. In the case of birds and mammals, the feathered and furry kind of species that people care about most, our knowledge is more accurate; these are estimated to number some 10,000 and 5,000 respectively.

The loss of some valued species is well documented: for example, the rapid disappearance of big fish. Since industrialized fishing began, declines of large predators in coastal regions have extended throughout the global oceans and 'large predatory fish biomass today is only about 10 per cent of pre-industrial levels'.[46] This makes a strong case for international action to control industrialized fishing, but it does not of course prove that environmental changes are causing the extinction of big fish. However, a disturbing picture emerges from the Red List of the World Conservation Union. This is a list of species evaluated by more than 500 scientists worldwide for threatened status, which comprises three categories: critically endangered, endangered, and vulnerable. The list almost completely covers the number of known mammals and birds and the latest list, for 2003, shows that a large number of species are threatened, although numbers have remained fairly constant in recent years. Out of some 5000 mammals, at the last count 184 were 'critically endangered', 337 'endangered', and 609 'vulnerable'. Out of some 10,000 birds, the categories were 182, 331, and 681 respectively.[47]

Over the past century, extinctions in well-studied groups, primarily birds and mammals, appear to average about one species a year, which experts tell us equates to a rate of about one hundred to one thousand times faster than the 'background rate'[48] (the rate observed previously). This sounds extremely high, and the estimate from the UN Global Biodiversity Assessment is even higher, namely 1500 times the background rate. Yet even the higher figure amounts to a loss of only 0.7 per cent of all animals, not per year but over the next fifty years.[49] While that may still be serious, it hardly justifies describing it as 'the opening stage of a human-caused biotic holocaust' as Myers has called it.[50] It should also be noted that all these forecasts are subject to a huge element of uncertainty.

One of the studies that gives serious grounds for concern about the effects of global warming is based on what has been described as one of the few iron-clad laws of ecology, the species-area relationship first demonstrated by Darwin's contemporary, H. C. Watson, which states that smaller areas support fewer species. This study suggests that a rise in world temperatures could have a major effect on the survival of various species. It assesses extinction risks for sample regions from South America, Africa, and Australia, that cover some 20 per cent of the earth's terrestrial surface. Its conclusion is that in the next fifty years the effect of climate change on the areas studied would cause the extinction of 18 per cent of species if the temperature rises 0.8 to 1.7 degrees Celsius, 24 per cent if it rises 1.8 to 2 degrees and 35 per cent if it rises more than 2 degrees.[51] Obviously the actual rate of global warming will be a crucial factor.

The study was based on a model and, as the authors acknowledge, is subject to many uncertainties. Computer models have been invoked to add an aura of objectivity to many misleading forecasts since the days of *Limits to growth*. I believe predictions based on models should be treated with a pinch of salt, even if I do not go so far as the US Federal Reserve Board chairman Alan Greenspan's comment about forecast movements in exchange rates: 'No model is superior to tossing a coin'. Nevertheless the study suggests that the risk of serious damage to biodiversity from climate change cannot be ignored.

It is frequently argued that mass extinction will result mainly from the loss of habitat that is being caused by deforestation, especially as a result of 'slash and burn' tactics carried out by local farmers. Yet an effective policy to prevent further deforestation is not an impossible, utopian dream. There is no reason why the international community should not act effectively. Conservation measures can make a difference, as they have to crocodiles, alligators, and caimans, most of which have now been moved out of the threatened category in the Red List. Furthermore, science can provide a variety of tools, including the genetic modification of crops, which make it possible to use land more efficiently and thereby avoid the need for poor farmers to destroy more forest. It can also help positively to increase biodiversity. No-till agriculture can have a beneficial effect on the quality of soil and the insects that live in it, which in turn can benefit the animals that feed on them. As mentioned, the Broom's Barn experiment demonstrated that herbicide-tolerant sugar beet can be managed in a way that increases the insect population on which birds depend (see p. 102 above). Other experiments have shown that conventional farming can encourage wildlife if managed appropriately. Biotechnology reduces the use of chemicals. All these developments suggest that practices that have done most damage to the environment in recent decades can be reversed.

Doomsters and distortion

Whether global warming is happening more slowly or quickly, whether the world's population will stabilize or continue to grow, whether we can feed the world with or without an environmental catastrophe, and whether or not there is likely to be a mass extinction of species, it is hard to see how any reasonable person can dismiss Lomborg's book or his detailed answers to its critics as lacking substance, or to see why he should be singled out for a special charge of scientific fraud.

I believe he has rendered a great service. Individual professional environmentalists may not subscribe to every article in 'the litany'. Perhaps Lomborg unfairly implied that environmentalists as a tribe share the views of veteran doomsters such as Paul Ehrlich and Lester Brown of the Worldwatch Institute, who have perennially exaggerated the problem of the world food supply.[52] However, these were not just men of straw for anyone to knock down. Their credo was adopted in many quarters of the Green movement, and a belief in coming catastrophe is still widely held today. It is the common coinage of widely read magazines like *The Ecologist* and is well documented in *Rising tides*. Since the publication of Lomborg's book there has been a perceptible change in the public pronouncements of a number of environmentalists. Before, it was generally assumed without much question that the environment was getting worse. Some rivers may be cleaner; so is the quality of air in London (and many other cities). But it was widely believed that the number of people dying from disease or starvation was increasing, that the supply of water was drying up, and that increasing consumption was leading inexorably to environmental degradation. Apart from global warming, the hole in the ozone layer was also seen as a source of impending disaster. Now it is much more frequently admitted that things may be getting better, that fewer people are starving, that even in the developed world people are living longer (except in places devastated by AIDS and many parts of Africa) and that rivers and seas may be cleaner than they were. The new message is . that, perhaps so far so good, but soon everything will be getting worse.

More than this, exaggeration of doom is often deliberate policy. Lomborg quotes Stephen Schneider, one of his principal accusers: 'Because we are not just scientists but human beings as well . . . we need to get some broad-based support, to capture the public imagination. That of course means getting loads of media coverage. So we have to offer up scary scenarios, make simplified dramatic statements, and make little mention of any doubts we have. Each of us has to decide the right balance between being effective and being honest'.[53] (The well-known quotation is somewhat unfair to Schneider, who added another sentence that is often omitted: 'I hope that means doing both.') Still, it is no wonder that the media constantly quote the higher figures for the range of global warming offered by the International Panel on Climate

Change. Surprisingly, Adair Turner condones tactics of exaggeration:

... green lobby groups operate in a market for public attention, in competition with companies selling their wares and business ... and if Green lobbyists eschewed the techniques of emotional appeal and, yes, sometimes slanted presentation, which companies and business lobbyists use, they would be unfairly disadvantaged.[54]

Lobbyists cannot be blamed for using emotional appeals, but Greenpeace, Friends of the Earth, the World Wildlife Fund, and the Soil Association are hardly a small band of powerless protesters desperately trying to make their voices heard and labouring at a huge disadvantage in the market place. They are multinational organizations with large memberships and vast resources. The annual income of Greenpeace International is over £100 million and that of the Soil Association over £6 million. Green organizations claim that their combined membership in the UK alone is over 5 million, although I suspect there is a lot of double counting.[55] Organic farming is now a billion-pound industry. Those who speak for green organizations can also count on wide coverage for anything they say from their many sympathizers in the media. They are far more skilful at using the media than the multinational companies and command a large army of volunteers. In the GM debate in Europe, it is the pro-GM voices that have to struggle to be heard. The anti-GM lobbyists were powerful enough to force GM products off supermarket shelves, despite no evidence of danger to health. A lobby group which, in the *Brent Spar* case, could compel Shell, one of the biggest companies in the world, to abandon a course that Shell was convinced was right (as indeed it proved), is not a weakling desperately trying to win attention against a wall of media hostility.[56] Ignoring facts, slanted presentation, or the repetition of allegations which are widely regarded as having been convincingly disproved are not minor peccadilloes that can be readily excused, but a major indictment of the way Green lobbies operate.

Eco-fundamentalists, like many other ideologues, believe that the end justifies the means. Distortion is justified and language is deliberately abused to nurture fears and create prejudice. For example, Greenpeace, Friends of the Earth, and the lobbies in favour of organic farming distributed a memorandum on the internet on how to manipulate the media. Its author, an organic industry advertising executive, advocated that accurate, science-based words were to be avoided—'biology', 'biotechnology', and 'food scientists', he said, 'are words we should never use'(since public reactions were less unfavourable). 'Make them use our words. Look how successful the term "terminator" seeds was. And congratulations on the success of the term "Frankenstein Food" '. The glossary of recommended terms was 'genetic engineering industry', 'genetically engineered foods', 'Frankenfood', 'test-tube food' and 'mutated food'. (The fact that organic farmers rely heavily on seeds mutated by irradiation was not mentioned.) Food produced by other than organic farming was to be referred to as 'non-organic', 'chemical-laden', and even 'toxin-laden'.[57] Indeed, it is regarded as axiomatic that: 'we have to offer scary scenarios . . . and suppress any doubts we have'. Truth does not matter if the cause is just. What matters is propaganda that achieves the right result and promotes the onward march of the true faith. That has been the cry of the authoritarian and the fanatic throughout the ages. Democrats take heed.

However, the threat to our society from eco-fundamentalists goes beyond indifference to evidence and distortion of language. Anti-scientific attitudes are gaining ground in Europe, as shown by the rising popularity of alternative medicine and organic food and of hostility to GM products. Europe is also gradually losing its scientific elite. Every year thousands go to study in the United States and most of them, more than 70 per cent, stay there because scientific careers and ingenuity seem more highly valued than in Europe. There are now some 400,000 European scientists based in America. Suppose green crusaders succeed in driving all bio-technology out of Europe, not only agricultural biotechnology which has already emigrated. They will not be content with victory over one enemy. Their rejection of technology goes much deeper and wider. Nanotechnologies may well be their next target.

Indeed they are inclined to categorize almost any technology promoted by multinational companies as a threat to man's union with nature. It is not inconceivable that triumphant green activism could cripple the future of technological innovation in Europe. Let us not forget the history of mediaeval Islam and of fifteenth-century China: Islamic ideologues froze progress in science in the Arab world when they led the world in science, and a faction in China banned shipbuilding, then the key to world influence and economic prosperity, when Chinese ships were the most advanced in all the world.

We are still a long way from such disaster. It is unlikely that Europe will allow green ideology to triumph. But this requires us to recognize the nature of the threat and to ensure that the forces of reason will prevail over the rising tide of eco-fundamentalism.

7

The Perils of Precaution

And always keep-a-hold of nurse
For fear of finding something worse

Hilaire Belloc

NOTHING more clearly demonstrates the change of mood from optimism and confidence about science at the time of the Enlightenment to present-day pessimism and suspicion than the current attitude to risk. It is a paradox of Western society that as people live longer and are less likely to die young from accident or disease, they become more conscious of risk. Indeed worries about possible harm seem to vary in inverse proportion to the probability of its occurrence. Parents are profoundly worried about the safety of their children—the hazards of vaccination, possible accidents, and possible murder by strangers. The fact is that a hundred years ago 15 out of 100 babies in England died before they were a year old. Now, as a result of vaccination, smallpox is extinct and diseases like measles and whooping cough are extremely rare in the developed world. Accidents in the home have declined substantially, as central heating has replaced open fires and gas fires, and electric lights have replaced candles. In Britain the number of children murdered by strangers is less than about one out of a million a year. People also worry about pesticides in food and about being killed in railway accidents. They are much less worried about the much greater risks of smoking, bad diet, or being killed when travelling by motor car.

According to a recent table published in the *British Medical Journal*, for every one of us the chance that we will not survive the next year is about 1 in a 100. There is a 1 in 7000 chance that we will die from an accident in our home, but only a 1 in 100,000 chance that we will be murdered. It is twice as likely that our home will be hit by a crashing aeroplane (a probability of 1 in 250,000) as that we will die in a rail accident (1 in 500,000). We are as likely to be struck by lightning as to die from CJD or in a nuclear power accident (a chance of 1 in 10 million).[1]

Risks that are unfamiliar, invisible, potentially catastrophic (however remote), or undertaken involuntarily seem particularly scary. In many cases there is an obvious psychological explanation: people are more ready to accept a risk if they have a choice or feel they themselves are in control, but are unhappy if there is nothing they can do to prevent an accident or illness. However, the idea put forward by some sociologists that 'the public understands uncertainty and risk well'[2] is wholly inconsistent with their behaviour in daily life. If the statement were true, the national lottery would not survive a week. In fact the public, or the popular press for that matter, have about as much understanding of the statistics of risk as of quantum mechanics. In particular there is a failure to distinguish between hazard, the potential harm in question, and risk, the chance that it will actually happen.

The Precautionary Principle

Clearly, society has a duty to protect its citizens against harm and it is not therefore surprising that when a new product is introduced and submitted for regulatory approval it undergoes a standard procedure of rigorous risk assessment. This applies to every genetically modified plant, every pharmaceutical drug, every new aeroplane, every new bridge. Scientists will calculate different probabilities of harm according to the nature of the product being tested. However, our current obsession with risk has introduced a new element into public life, which was originally called the precautionary approach but has now been elevated into the so-called Precautionary Principle (with capital letters). It is now so widely accepted it has come to be regarded almost as an eleventh commandment: 'Thou shalt not take any unnecessary risk', or as a new scientific law. For decades the *Vorsorgeprinzip*, or foresight principle, has been incorporated into German environmental law. In 1982 it was recognized in the World Charter of Nature ratified by the United Nations General Assembly, it was formally adopted by members of the European Union in the Treaty of Maastricht, and it is constantly invoked by government spokesmen in the UK. Thus in 2002, when the British Prime Minister, Tony Blair, delivered a major speech on science (which was generally well received by scientists), he said: 'Responsible science and responsible policymaking operate on the Precautionary Principle'.[3] It is one of the guiding principles of the Green lobbies, who invoke it frequently to stop scientific developments they oppose, of which there are many and of which GM crops are the most prominent example. It raises a sympathetic public response. What could be more sensible than a policy of 'Better safe than sorry'?

Given its wide acceptance, we might have expected the principle to have been clearly defined, but this is not the case. There are at least fourteen official definitions, none of them particularly helpful and some positively harmful. One frequently quoted formulation appeared in the principles adopted by world leaders at the Rio Conference on the Environment in 1992: 'Where there are threats of serious or irreversible damage, lack of full scientific certainty shall not be used as a reason for postponing cost-effective measures to prevent environmental degradation.' At first sight, this definition seems to be quite reasonable, if not rather obvious. Of course, we must exercise caution if there is a real risk of serious damage. But in fact the definition is so vague it is useless. What constitutes a threat? What is serious damage? How serious must it be? How do we know that damage will be irreversible? All sorts of diseases that were incurable can now be cured. Rachel Carson forecast that the nitrification of the Great Lakes was irreversible, yet the damage has been reversed. Again, there is never, as Hume showed, full certainty in science. To apply the Rio definition would enable almost anyone to invoke the principle at any time on almost any grounds.

Another wordy version of the principle is to be found in the Cartagena Protocol on Biosafety, 2000, signed by leading Governments:

in accordance with the precautionary approach the objective of this Protocol is to contribute to ensuring an adequate level of protection in the field of the safe transfer, handling and use of living modified organisms resulting from modern biotechnology that may have adverse effects on the conservation and sustainable use of biological diversity, taking into account risks to human health, and specifically focussing on trans-boundary movements.

This says nothing at all. But the most common definition used is that 'when an activity raises threats of harm, measures should be taken even if some cause and effect relationships are not established scientifically'.[4] This version abandons any restraint and makes decisions about whether innovations that require a government licence should be permitted to depend on articles or letters in the press, campaigns by Green lobbies, and public fears and alarums. If scientific evidence is no longer the test, caprice, not reason, reigns supreme.

Nevertheless, legislators and Governments take the principle seriously. In 2002 the European Environment Agency (EEA) published a report on the Precautionary Principle, or more specifically on the 'use, neglect and possible misuse of the concept of precaution in dealing with a selection of occupational, public and environmental hazards' over the 100 years between 1898 and 1998.[5] It presents an analysis of past mistakes and is perhaps the most thorough attempt yet made to suggest guidelines for the future use of the Principle. However, this report on precaution should itself be treated with great caution, because it is selective in its use of evidence and demonstrates a marked bias, not uncommon in environmentalist circles, towards a pessimistic interpretation of events.

Firstly, all the fourteen cases selected for examination are cases of unexpected harmful consequences from some new product or process, suggesting that uncertainty always results in harm. Had the authors been concerned to present a balanced picture, they would also have discussed cases, of which there are many, where innovation was not followed by harm but by unforeseen benefits. No whiff of optimism here. Furthermore, the authors are unaware of serious studies that do not support their findings, of which they should have been aware. Alternatively they chose to ignore them. For example, when discussing a study by an epidemiologist which found that the use of X-rays on pregnant mothers caused leukaemia in their children, the authors state: 'A similar and contemporary story may be unfolding in relation to the childhood leukaemia risk in proximity to overhead power lines in the United States'.[6] In fact the most exhaustive epidemiological studies ever undertaken, published two years before the EEA report, found no evidence of such a link (see pp 177–178 below).

The worst example of one-sided reporting is to be found in a case study of chemical contamination of the Great Lakes.[7] The report refers to claims made in 1978 by people living near Love Canal, Niagara Falls, New York, that high rates of birth defects, miscarriages, cancers, and other health problems were caused by the leakage of toxic chemicals into the canal. They lived in a community of about a thousand homes, some of which were built on top of an old chemical waste tip. Dioxins and other chemicals seeped from this tip. The US Environmental Protection Agency (EPA) became worried about possible genetic damage and other health risks from the leaked chemicals and reported that there was real danger. Love Canal was accordingly declared an emergency area, the families were evacuated, the canal was dredged, and the contaminated sediment was sealed in drums and taken away. The incident has been much celebrated as a successful campaign by local action groups and in 1990 its leader, a housewife called Lois Gibbs, received a prize for her work from the Golden Environmental Foundation of San Francisco.

The inference we are supposed to draw from the EEA report is clear, that the harmful effects of the dioxins were proved, although this was consistently contested by local, State, and government officials. Inexcusably, there is no mention of a later study that did not confirm the claims that dioxins had caused the illnesses of which the residents complained. The Centers for Disease Control, alarmed by what had happened, carried out a survey in which the health of the Love Canal residents was compared with that of a similar community far removed from the area. The analysis was conducted on a double-blind basis to eliminate subconscious factors and the survey found 'that the illnesses afflicting the residents of Love Canal were not unusual, but were to be expected in a normal community of that size'.[8]

Despite bias towards pessimism and the selectivity of the examples chosen, the case studies discussed in the EEA report provide much useful information. The studies fall into three categories: (a) those in which there were clear warnings of danger which were ignored or not taken seriously; (b) those in which the extent of the hazard became evident gradually over time; and (c) those in which the consequences were unforeseen. The history of asbestos, of radiation and radio-activity, and of chlorofluorocarbons (CFCs) are examples from each category.

(a) The story of asbestos

The story of asbestos is one of the worst examples of the failure of Governments to heed clear warnings. Mining for asbestos started in 1879, when it was regarded as a mineral that would greatly benefit mankind because it could save lives by preventing the spread of fires. There is no doubt that it did so. However, within twenty years Lucy Deane, one of the first female Inspectors of Factories in Britain, drew attention to its dangerous effect on the health of workers through damage to their bronchial tubes and lungs. She reported that microscopic examination of the mineral dust had 'clearly revealed ... the sharp glass-like jagged nature of the particles, and where they are allowed to rise and remain suspended in the air in the room in any quantity, the effects have been found to be injurious, as might have been expected'.[9]

In 1909 and 1910 similar warnings were widely circulated

among policy makers and politicians but were simply ignored. By the 1920s, insurance companies in the United States and Canada knew enough about the dangers of asbestosis to refuse insurance cover. Only in 1931 were the first regulations for prevention and compensation introduced in Britain and even then they were not rigorously enforced. When evidence came to light in the mid 1960s that asbestos caused mesothelioma, usually a rare cancer of the lining of the chest and abdomen, effective action was again delayed. The Dutch Ministry of Health estimated that had a ban been introduced in 1965 instead of 1993, 34,000 cases of mesothelioma would have been prevented in the Netherlands alone. Indeed because of the long latent period, many more thousands of people will continue to die of mesothelioma until the year 2030.[10] The history of asbestos is one of scandalous failure to take notice of clear evidence of harm that has cost, and will continue to cost, innumerable lives.

(b) X-rays and radiation

X-rays were discovered by Wilhelm Konrad Röntgen in 1895. He recognized their value for medical diagnosis and immediately published his findings. There were some early warnings of the harm they could do, but it was only as evidence of damage from exposure to X-rays gradually accumulated that the first steps towards protection were taken, by the German Radiological Society in 1913. A few weeks after the publication of Röntgen's work, Henri Becquerel discovered radioactivity, and then in 1898 Marie and Pierre Curie discovered radium. Both Becquerel and Pierre Curie suffered skin erythemas from carrying lumps of radioactive materials in their pockets, Pierre Curie died from exposure to radiation, and Marie Curie died from leukaemia, which may well also have been induced by exposure. But the true risks from radioactivity were only gradually recognized. For example, a third of the young women working in factories making clocks in the First World War and throughout the 1920s, who used to lick their brushes into a point before applying radioactive luminous paint to the dials, eventually died of various malignancies. In 1928, an International X-ray and Radium Protection Committee was formed, and slowly standards of protection were evolved, in each case after some time lag. The EEA report remarks that there have always been periods when changes in standards have lagged some years behind clear evidence of harm to human health.[11]

It should be mentioned that the harmful effect of any dose of radiation on human health has been overstated. Present policy is based on the 'linear no-threshold' assumption endorsed by the International Radiation Protection Board. This assumes that even the smallest dose of radiation is harmful and may cause cancer and genetic disorders. Unfortunately, far from safeguarding us, current safety standards may in fact result in an increase in the incidence of cancer.[12] There is evidence that the harmful effect of radiation is not linear, but that low doses may actually be beneficial through the hormesis effect (see Chapter 3, pp 72–3). Paradoxically a low dose of ionizing radiation may stimulate DNA repair and some immune responses, thus providing a measure of protection against the development of cancer. The benefits of ionizing radiation in treating cancer are well known, but that general exposure to low doses is beneficial rather than harmful is confirmed by a mass of evidence, particularly from Japan, where the long-term effects of radiation on the population have been studied in the areas around Hiroshima and Nagasaki. Furthermore, the death rate from leukaemia of workers in the nuclear industry in Canada is 68 per cent lower than the average for the population and workers in nuclear shipyards in the USA and many other countries have substantially lower death rates from all cancers. Indeed there is also clear evidence that people who live in areas of unusually high natural radiation in Japan, China, the United States, and India are less likely to die from cancer than control groups.[13]

(c) CFCs

One of the most interesting case studies highly relevant to current arguments about the Precautionary Principle is that of the effect of chlorofluorocarbons (CFCs) on the ozone layer. Industrial production of CFCs started in the 1930s and at first there was no reason to suspect that they might have any harmful effects. To quote the EEA report:

There can be no doubt that a conventional risk assessment, in say, 1965, would have concluded that there were no known grounds for concern. It would have noted that CFCs were safe to handle, being chemically very inert, non-flammable and having very low levels of toxicity.[14]

They had been released into the atmosphere for thirty years with no apparent harm being done. In fact it was not until 1970 that concern about the effect of human activities on the ozone layer became an international issue. Initially fears were expressed about emission of nitrogen oxides, carbon monoxide, and water from supersonic aircraft. But then attention shifted to CFCs because it was found that as the CFC industry had expanded enormously during the 1960s, these gases—released largely into the atmosphere in the northern hemisphere—had spread around the world. In 1974 American scientists pointed out that CFCs were such stable gases that they would eventually reach the stratosphere, where it was likely that chlorine would be released by a reaction with light and an ozone-destroying chain reaction would ensue. In 1977, after a public campaign, legislation was passed in the United States, Canada, Norway, and Sweden banning CFCs as aerosol propellants. In Europe action was not taken until 1980 when a Council directive was passed which aimed to freeze the production of key CFCs and reduce their use in aerosol by 1981 by at least 30 per cent from 1976 levels.

However, by the early 1980s concern about the ozone layer had subsided. According to the EEA report, computational models predicted only small long-term reductions of ozone, and this was in reasonable accord with observations that showed no significant trend. Then, in May 1985 an article was published in the scientific journal *Nature* that reported rapid and severe ozone depletion over Antarctica, much more severe than any prediction, and this was confirmed by NASA in October 1985. Discovery of the ozone hole, as it was later called, came, as it were, out of the blue as a result of systematic long-term measurements begun solely for scientific exploration. This was not a case therefore in which Governments ignored clear warnings. The EEA report on the subject of the ozone layer concludes: 'It should not be assumed that environmental science has reached the stage where all hazards can be foreseen. All too often technology outstrips the science needed to assess the risks involved'.[15]

Concern about unforeseen hazards forms the core of one of the lessons drawn by the Agency about the Precautionary Principle. Many of its conclusions are sound, if obvious: they tell us that warnings should not be ignored, that all relevant evidence should be considered, that if there is clear evidence of serious harm we should not wait for full certainty before acting, that regulatory authorities should be independent, and so on. However, elsewhere the report also suggests that non-scientific evidence should be taken into account and that we should consider not only uncertainty but ignorance. In this it echoes a theme developed by a group of contemporary sociologists and environmentalists who are among the most prominent advocates of the Precautionary Principle and who berate policy makers for not involving lay opinion in the assessment of risk and for failing to take into account, not only 'unknowns' but also 'unknown unknowns'.[16]

Non-scientific evidence

The possibility that, despite the absence of evidence, application of the Principle may be triggered by press speculation or campaigns by lobby groups or unspecified public fears is not a theoretical one. One example comes from the United States where there was widespread concern that electro-magnetic fields (EMF) created by power lines cause cancer, particularly leukaemia in children. In 1989 Paul Brodeur, a writer for *The New Yorker*, described EMF as the most pervasive health hazard to Americans, which he alleged had been concealed from the public by a conspiracy.[17] There was some rather weak evidence of an unusually high incidence of leukaemia among children living near power lines. Naturally parents were deeply worried and for some years press reports assumed that a link existed. Epidemiological studies were commissioned, but they found no such link. Their reports were invariably followed by demands for further study or by insistence that the absence of a link should be positively proved. There were also calls for a policy of 'prudent avoidance' and claims of a cover-up. Only after the most extensive epidemiological studies ever undertaken on any one subject, at a cost of over US$25 billion incurred through the relocation of power lines and the loss of property values, was it finally accepted in 1996 that there is no relationship between EMF and leukaemia.[18]

In Britain there were similar claims, mainly raised in the press, that EMF emanating from mobile phones and mobile phone masts caused a number of illnesses.[19] There were stories that 'mobile phones cook your brain' (*Sunday Times* 4 April 1996) and cause hypertension, miscarriages, and loss of memory, that phone masts cause cancer, and that 'sickly pupils recover after leaving phone mast school' (*Daily Express*). While some parents of children at schools situated near phone masts did express concern, most complaints from the public were about the ugliness or siting of the masts. Mobile phones became ever more popular. Nevertheless, the Government set up an inquiry in 1999 (the Stewart Inquiry), not because there was scientific evidence or great public disquiet, but in response to the press campaign and, in the words of the Health Minister at the time, 'to keep ahead of public anxiety'.[20] In due course, the inquiry reported that there was no evidence of harmful effects, although it also recommended that, for extra security children should use them as little as possible. Children and their parents ignore the advice; sales of mobile phones continue to rise; campaigns by parents against mobile phone masts have carried on as before.

Rule by scare story panders to, and reinforces, the prevailing mistrust of experts. It flatters public opinion by suggesting that the instincts of the public are basically sound. In fact it strengthens public concerns, because when the Government reacts by setting up an inquiry it is argued that the Government would not have done so unless there was something substantial to worry about. Yet however democratic it may be to assume that somehow the public *en masse* has access to sources of knowledge or possesses some innate wisdom that remains closed to experts, it is hardly a sound basis for making policy. The instincts of the public may sometimes be right and sometimes be wrong. Had the fears of the public dictated public policy in the past, Jenner's smallpox vaccine would not have been developed. Today we would have three separate vaccines in place of the combined measles, mumps, and rubella (MMR) vaccine because of public fear that it causes autism, and a serious measles epidemic would be even more likely than it is today in the aftermath of the anti-MMR campaign (see Chapter 2, p. 50).

Allowing for 'unknown unknowns'

Perhaps the most important and controversial recommendation of the EEA report is that we should take account of uncertainty and ignorance. Of course there is never certainty in science and regulators have to deal with uncertainty all the time. As already mentioned, when a new drug or food or other technological development has to be licensed, the authorities have to be satisfied that it has passed appropriate safety tests. The kind of proof required will vary according to the innovation. Sometimes we may need proof 'beyond all reasonable doubt', sometimes 'a balance of probability', sometimes 'reasonable grounds for concern' may be enough to refuse a licence and sometimes (if the hazard to be guarded against is a very serious one), the authorities act even if there is only 'a suspicion of scientific doubt'. It all depends on the circumstances. However, the EEA report invites us to 'respond to ignorance, as well as uncertainty'.

Perhaps not surprisingly, there is a lack of clarity in this section of the report and a frequent resort to vague generalizations. We are informed, for example, that we need multi-criteria mapping, which 'combines the flexibility and scope of qualitative approaches with the transparency and specificity of quantitative disciplines' and that 'the Precautionary Principle has nothing to do with anti-science, and everything to do with the rejection of reductionist, closed and arbitrarily narrow science in favour of sounder, more rigorous and more robust science'.[21] There is no indication how this 'more rigorous and robust science', whatever it may be, can help us allow for ignorance or 'unknown unknowns', or how we can guard not only against the perils we know about or which we have some grounds to suspect are threatening us, but also against perils we do not know about and when there is no reason to suspect a threat. Flexible 'qualitative approaches' presumably differ from 'reductionist science' in that they are not susceptible to the pedestrian process of testing evidence, conducting double-blind tests, or changing one variable at a time to determine cause and effect. It signals a rejection of the scientific method and the evidence-based approach in favour of intuition, or just plain sloppy irrationality.

Holger Hoffman-Riem and Brian Wynne tried to explain in the correspondence column of *Nature* why we have to take account of ignorance.[22] It is worth quoting this letter at some length, because the concept of 'unknowns' is frequently invoked and because Professor Wynne is an influential academic, who was, for example, special adviser to a House of Lords Select Committee on Science and Society in 2000 that produced an important report recommending a more 'democratic' approach to science. Scientific knowledge, the authors tell us,

refers to known processes and their influence upon known state-variables. . . . The domain of ignorance is characterized by the interaction between unknown processes and/or unknown state-variables, [which] tends to be implicitly neglected in risk assessment.

They then cite the examples of DDT and CFCs. In the case of DDT

we are dealing with the interaction between a known process (increase in DDT concentration) and an unknown, thus neglected, state-variable (egg shell thickness); between this neglected state-variable and a neglected process (population dynamics); and between this neglected process and the neglected state-variable of bird population. All of these interactions fell within the domain of ignorance for the contemporary risk assessments of DDT. . . . In the case of CFCs . . . their concentration in the stratosphere was not monitored (neglected state-variable). No one suspected a connection between stratospheric CFC concentrations and stratospheric ozone concentrations (neglected photochemical process in the stratosphere). Once again these state-variables and interactions were neglected because they belonged to the contemporary domain of ignorance.

What this letter says, dressed up in somewhat pretentious language, is simply that some of the effects of DDT and CFCs were not foreseen. And because they were not foreseen, they were not taken into account. From this breathtaking revelation, they conclude that in future 'Multiple interacting perspectives should be encouraged. Lay knowledge in particular can be a valuable addition to expert knowledge, because it is based on different experiences.' Unsurprisingly, it is not made clear how lay knowledge would have helped foresee the effect of DDT on eggshell thickness or of CFCs on the ozone layer. In fact the hole in the ozone layer (and the existence of the layer itself) was discovered by high science, not by laymen or Children of the Celtic Dawn.[23]

Nevertheless there is a question to be answered: since the effect of CFCs on the ozone layer was not foreseen, should we perhaps be more cautious than we have been? The public, and many advocates of the Precautionary Principle, always come back to the traumatic experience of BSE. If we do not know whether something will cause harm, would it not be wiser not to take a chance? The answer is that every innovation has always had unforeseen consequences, many of them harmful. If we aim to avoid all actions that might conceivably cause harm, we would do nothing. Even if allowance were made for benefit in those cases where clear benefit could be foreseen to weigh against unknown dangers, many of the most notable inventions of the past, from life-saving drugs to aeroplanes, would not have survived in the new climate of suspicion and distrust that is obsessed with unknown dangers. We would have to stop the world because some people want to get off.

Does that mean we must accept that there may be future BSEs? The first answer is that it is not humanly possible to guarantee that future disasters will not happen. Secondly, it is important to identify the real lessons to be learnt from the BSE episode. There was a degree of ignorance at the time about the cause of the outbreak of spongiform encephalopathy in cattle (which is still obscure) and there was an assumption, perhaps too readily made, that BSE was a form of scrapie, a prion disease of sheep, which had never been transmitted to humans. However, contrary to popular belief, scientists did not guarantee the public that eating beef was safe. Despite the limited knowledge of spongiform encephalopathy in government circles and the mistaken assumption about scrapie, what scientists said was that the risk of humans being infected was 'small'. In retrospect, that view seems likely to have been correct. The guarantee of safety came from politicians and civil servants, who were worried (not unreasonably) about causing panic and did not accurately pass on the advice they received. The major errors were a lack of openness and a basic conflict of interest inside the old Ministry of Agriculture, Food and Fisheries, which was responsible for both food safety and farmers' interests and which allowed the interests of the latter to prevail.[24] The public is entitled to expect that such errors, especially the obsession with secrecy inside government, are not repeated. However, failure to apply the Precautionary Principle was not the cause of the spread of the infection. And there are dangers in following the seemingly innocuous recommendation of the BSE Inquiry report that 'The importance of precautionary measures should not be played down on the grounds that the risk is unproved'.[25]

The perils of over-precaution

The more closely the Precautionary Principle is examined, the less useful and indeed more harmful it is found to be. What the EEA report demonstrates is that attempts to define it are vain because definitions are either obvious ('Be careful if there is evidence of danger') or absurd ('Guard against the unknown'). The main use of the principle is to justify opposition to new technology, which is why it is so popular with opponents of the genetic modification of plants who invoke it in favour of a ban or moratorium on their planting for commercial use. What is less immediately obvious are three other major defects: the principle is used to ban products and substances which present no serious risk of damage to human health; no regard is paid to the need to strike a balance between benefit and risk, and no allowance is made for unforeseen benefits.

Exaggerated and non-existent risks

Unjustified fears about the effects of low doses of radiation have already been mentioned. Another case that clearly illustrates the absurdities arising from over-strict application of the Precautionary Principle is that of the ban on the importation into Europe of shrimps and other fish products from Asian countries imposed in 2001 because they were found to contain the broad spectrum antibiotic, chloramphenicol.

Chloramphenicol is one of a number of pharmacologically active substances that are treated by EU regulations as 'dangerous at any dose'. Under the Precautionary Principle as applied to food safety issues, zero tolerance is prescribed: 'When in doubt, keep it out'. It is categorized by the International Agency for Research on Cancer as 'probably carcinogenic in humans'. Its most dangerous side-effect, however, is its association with aplastic anaemia, an extremely rare but fatal condition in which the bone marrow ceases to produce red blood cells. Apart from topical use in eye drops for eye infections, its use is now largely confined to the treatment of bacterial meningitis, typhoid fever, and, specifically, typhus, and the risk of developing aplastic anaemia is very low, at less than 1 per 1,000,000 treatment regimens.[26]

There has never been a case of aplastic anaemia as a result of exposure to chloramphenicol in food. Its presence was only detected in imported shrimps because modern methods of detection are so sensitive that compared with the previous threshold of one part per million, they can now detect one part per thousand million. It was assumed that its presence resulted from illicit veterinary use in the countries of origin, although it could easily have come from other sources as it is a natural product of organisms commonly found in the soil. The lowest concentration of chloramphenicol to have a pathological effect is not known. As evidence accumulates in support of the hormesis effect, suggesting that small amounts of many compounds may in fact be beneficial (see Chapter 3, pp 72–3), the whole basis of the doctrine of zero tolerance is undermined. To invoke Paracelsus once again, it all depends on the dose.

The ban therefore defied all reason. It was positively harmful to the poor: affluent European consumers imposed a ban at the expense of the livelihood of Third World fishermen and shrimp farmers. It is another example of chemophobia, the pressure to ban or regulate any chemicals that might conceivably have any harmful effects that has been the driving force behind the REACH (Registration, Evaluation and Authorisation of Chemical Substances) programme of the European Union a programme based more on unjustified fear than rigorous regard for evidence.

Balancing benefit against risk: the case of DDT

DDT is perhaps the most outstanding example of the failure of the EEA report, and of the Precautionary Principle itself, to balance harm against actual benefit. Since Rachel Carson wrote *The silent spring*, it has been a common assumption that DDT was another technological disaster and should have been banned long before public spraying for agricultural purposes ceased. Typically, the report uncritically accepts the conventional wisdom and assumes that all use of DDT should be eliminated. In fact the story of DDT shows how the values of many environmentalists have become distorted and have ceased to strike any reasonable balance between concern for animal life and concern for humankind.

Rachel Carson may have been an inspiration to all who cared about nature, but she overstated her case against DDT, not only in relation to its effect on the bald eagle and peregrine falcon (see Chapter 1, p. 29). She claimed that it caused cancer of the liver, and cited anecdotal evidence of other sorts of severe damage to health. Yet not a single study showing that exposure to DDT damages the health of human beings has ever been replicated.

On the other side of the balance sheet as mentioned is the fact that DDT is one of the most effective means of preventing disease ever conceived, the most effective method of killing mosquitoes and preventing the transmission of malaria. By the early 1960s its use had virtually eliminated malaria from southern Europe, the Caribbean and parts of east and south-east Asia. In Sri Lanka there were only 31 cases of malaria in 1962, 17 in 1963, but more than a million cases in 1968 after the use of DDT was banned.[27] In 1970 the US National Academy of Sciences reported that 'in a little more than two decades, DDT has prevented 50 million human deaths from malaria'. Similar statements were made by the WHO.[28] The life-saving properties of DDT are not minor benefits outweighed in the scales by the threat to wildlife: DDT has been of massive benefit to mankind. Since its use was effectively banned, the number of cases of malaria has increased dramatically and it now kills over a million people a year, mainly children.

No one now advocates a return to the use of DDT for agricultural spraying. Today, there are better ways of protecting crops against pests, such as by the genetic modification of crops to make them pest-resistant. While the beneficial effects of DDT spraying in the fields were only temporary, because in some areas mosquitoes developed resistance to it, when sprayed on to the inside walls of houses it acts as a repellant and irritant and either kills or drives away mosquitoes before they have a chance to bite. Mosquito nets dipped in pyrethroid are effective, but eco-friendly approaches including mosquito-repellent trees or fish that eat mosquito larvae have proved much less useful. In the mid 1990s, South Africa switched from using DDT to using pyrethroid, but found that the number of cases of malaria greatly increased as mosquitoes became pyrethroid-resistant and it therefore returned to the use of DDT which reduced the incidence of malaria by 75 per cent in two years. DDT is more effective and cheaper and has to be sprayed less often than pyrethroids. Yet the aid agencies refuse to fund DDT programmes and countries like Uganda have been told that Europe and the United States might ban their fish and agricultural exports if DDT is used.[29] So far the story of the reaction against DDT is one of the triumph of the politics of 'saving the planet' over the science of preventing disease. It is the obverse of the asbestos scandal. In that case failure to act caused thousands of deaths. In the case of DDT, overreaction was responsible for some millions of deaths. If a total ban on DDT worldwide is made effective, it will be a victory for the conscience of the rich world, invoked without regard for facts, at the expense of the lives of the inarticulate poor.

Unforeseen benefits

In its failure to envisage unexpected benefits, the EEA report sides with the doomsters and eco-fundamentalists and gives us a skewed picture of the proper role of caution. Every new invention is bound to have consequences that are in 'the domain of ignorance', but the odds are that they are as likely to be beneficial as harmful. Why should we only be concerned with the possibility of harm? No one foresaw that aspirin would turn out to be a wonder drug, not only useful as a painkiller, but as an agent to stop blood clotting, inflammation, heart attacks, and strokes. Viagra was developed as a drug to cure baldness and its more stimulating effects that have proved so popular were unexpected. Faraday, when he discovered electricity, said he saw no practical use for it, except that ways might be found of taxing it. When the optical laser was invented in 1960, it was dismissed as 'an invention looking for a job'.[30]

It is unlikely that aspirin would have passed the test of the Precautionary Principle, because adverse side-effects *were* known, while most of its benefits were wholly unexpected.

Perhaps the most startling example is the history of thalidomide. It was first prescribed in the late 1950s as an effective sedative, particularly useful to young children with serious brain disorders, but was then found to cause birth defects when given to pregnant women, a result which made it the most infamous drug in the world. Errors were made when the drug was licensed in Europe because it had not been properly tested on pregnant animals. Yet thalidomide is a good example of a drug that has since been found to have unforeseen beneficial properties and even greater potential. For example, it has been used for twenty-five years to treat *Erythema nodosum leprosum*, a nasty reaction in patients who have leprosy, and it has also proved remarkably effective in treating mouth ulcers in HIV-infected patients. It is currently undergoing trials for use in breast, prostate, lung, and renal cell cancers and has already been approved by the Food and Drugs Agency in the United States as a therapy for multiple myeloma. Indeed, it has been described as the first really effective therapy for myeloma in almost twenty years.[31] Who would have foreseen when it was withdrawn from the market in the early 1960s that within forty years it would promise such benefits?

There is at least one case in which the application of the principle had disastrous effects. In the early 1990s, the Peruvian Government stopped the chlorination of much of the country's drinking water, when it found itself subject to budget constraint, on the grounds that chlorination posed a potential cancer risk. Environmentalists have long campaigned against organochlorines on the grounds that they have carcinogenic effects, and on this occasion officials listened to them. The Government's decision helped accelerate and spread the cholera epidemic in Latin

America of 1991–6, which affected more than 1.3 million people and killed at least 11,000.[32]

In practice, the Precautionary Principle may often be applied sensibly. However, because it operates asymmetrically and emphasizes possible harm, not benefit, it is bound to tilt the balance against innovation. It happens today with the licensing of new drugs. A regulator who stops a new drug suspected of harmful side-effects is a hero or heroine. One who licenses a drug that would have clear benefits, though not without the possibility of doing some harm to a few people, would be considered irresponsible. It is considered to be an act of inexcusable negligence to allow a useful life-saving drug to be prescribed which may have side-effects that may contribute towards a hundred deaths. No blame is ever incurred for delaying licensing or banning the use of a drug that could save a million lives. In fact, we almost certainly kill more people by delaying the marketing of new drugs than we save by testing them exhaustively.[33]

The Principle is particularly favoured by those who wish to block controversial technology, such as the opponents of GM crops. They constantly make the impossible demand that GM crops be positively proved safe. There is of course no absolute certainty in science and the demand cannot be met. No food of any kind can be proved safe and the absence of evidence of harm after careful testing is the only possible basis for government policy. In fact the demand is made in order to stop the cultivation of GM crops altogether. Indeed, Martin Teitel, the leader of the American activist group, Council for Responsible Genetics, admitted that rather than make the politically difficult demand that the science of biotechnology should be shut down, 'requiring scientists to satisfy the Principle by proving a negative means that "they don't get to do it period".'[34]

The no-risk society

There is another reaction to ever-growing concern about safety that has overstepped the boundaries of reason, namely the multiplication of health and safety regulations. Of course, some regulations are not only necessary but highly desirable. One reason why apocalyptic predictions about the decline in the quality of our environment have proved wrong is that many Governments have very sensibly enacted laws and regulations to protect it. These have improved the quality of the air we breathe, the water we drink, and the beaches we swim from and have prevented accidents at work. But other regulations have been passed which make no attempt to balance cost against benefit and which seem designed not merely to reduce risks, but to eliminate them altogether. They overwhelm businesses, institutions of all kinds, doctors, teachers, and other professions with a mass of paperwork.

Like the Precautionary Principle, regulations to protect us from almost any kind of risk reflect this yearning for a no-risk society that is becoming an ingrained feature of contemporary culture. Bureaucrats have an interest in multiplying regulations because it is one way of protecting themselves from criticism. Many organizations and professions have a vested interest in exaggerating risk because it increases the demand for their services: firemen exaggerate the risk of fire, policemen of crime, lawyers benefit from a proliferation of laws. Our worry about the consequences of ignoring risk has even spawned a movement to reconsider the use of the word 'accident'. As most injuries can be prevented, it is argued that it is socially irresponsible to describe them as accidents. Indeed, in 2001, the *British Medical Journal* stated that it would no longer use the word, because even earthquakes, avalanches, and hurricanes were predictable events, which could therefore be avoided, or against the effects of which we can take necessary precautions. As one commentator observed, 'Some child professionals insist that we should refer to a youngster's bruised knee as a preventable injury, rather than an accident.'[35]

There are many extraordinary examples of claimants in the United States winning damages for defendants' failures to take extreme precautions against injury, the plaintiffs themselves being absolved from any personal responsibility. Obese plaintiffs are suing McDonald's for failing to warn them that too much fatty food could make them fat. I know one expatriate who has left the United States because he is no longer allowed to ski 'off-piste', since ski clubs fear they would be sued in the case of a mishap. In Britain, a soldier sued the Ministry of Defence because he was not warned about the horrors of war (he lost his case). These are not isolated examples of anecdotal evidence, but reflect a widespread characteristic of contemporary culture. Almost invariably the first question asked by a journalist who interviews someone involved in an accident is: 'Who do you blame?' In these circumstances, it is not surprising that regulations seek to protect governments and other public bodies from responsibility for any hazard they can conceivably envisage.

The irony of obsession with avoiding risk is that it has the opposite effect of what is intended. Measures to avoid risk make people more worried. After all, if governments go to such lengths to make detailed regulations to prevent something happening, there must be a good chance that it will happen. Why else would they bother? It should therefore come as no surprise that Sweden, the country that has led the way in making us aware of the supposed risks from the use of chemical substances, suffers a notably higher incidence of psychosomatic symptoms caused by fear of such risks than any other country.[36]

What would a no-risk society look like? The ultimate safeguard against playground accidents to children would be to abolish playgrounds. Children would be safe from pædophiles if they were not allowed out to play or to walk home from school by themselves. They would not drown if they never went swimming. Climbing accidents could be minimized by allowing only climbers who had a qualified licence to climb mountains. Crime would be drastically reduced if curfews were in force that restricted the right to go out at night to specially licensed citizens. In Britain and some other countries sailing is at present a largely uncontrolled activity: anyone can take a boat to sea, however unseaworthy the boat and however little the skipper knows about sailing. This absence of regulation and freedom to harm oneself, which is a joy to sailors, is clearly something the no-risk society would never tolerate. Indeed, in France, a licence is already required to take a yacht out to sea, and accidents caused by the seasickness of crew members have led to prosecutions for setting sail with an unseaworthy crew. Ireland has enacted a law that crew on sailing boats of less than seven metres must wear life-jackets at all times.

The worst effect of a no-risk society would be on the future of science and technology. Caution would be the watchword for any new invention: that in effect is what the Precautionary Principle says. It is the watchword of pessimism. It is the triumph of the Jeremiahs. It is the victory of the Spartan spirit, fearful of the terrors change may bring, over the Athenian spirit that looks for new worlds to conquer, and of the Luddites who want to stop innovation over those who want to try it out. The urge to cross new frontiers of knowledge, which science stands for, is something the no-risk society would seek to stifle at birth. A no-risk society would be a world without excitement, without exuberance, free spirit, imagination, or innovation, doomed to gradual economic and intellectual decline. It would be a paradise only for lawyers.

8
The Attack on Science

Truth is what your contemporaries allow you to get away with.

Richard Rorty, American philosopher

So far, the reactions against science and technology I have described have not been based on intellectual arguments, but mainly reflect a general instinctive malaise—that natural remedies and natural farming methods are best, that we are losing touch with nature, and that science is subjecting us to ever greater risks of harm. However, there has also been an intellectual assault on science in academe, a philosophical, political, and sociological critique whose impact on attitudes to science should not be underestimated. The assault has been launched under a number of banners, such as cultural relativism, deconstructionism (which has attacked science for 'reductionism'), and hermeneutics (literally, the art or science of interpretation), but I shall concentrate on ideas generally associated with relativism, or, in the particular case of the attack on science, postmodernism.

Postmodernism is not a theory or school of thought that is easy to define. Few philosophers today seem willing to describe themselves as postmodernists. Its main impact on the public has been as a movement in literature and the arts, but, although its authority has faded, its critique has been adopted by many teachers of sociology, mainly in the United States, and has left its mark on attitudes to science among generations of students, adding weight to the current mood of scepticism and suspicion. In particular, it has had an influence in turning against science those on the left of politics, who were traditionally supporters of the scientific approach. In the words of Professor Alan Ryan, 'American departments of literature, history and sociology contain large numbers of self-described leftists who have confused radical doubts about objectivity with political radicalism and are in a mess'.[1] Its main inspiration came from a group of French intellectuals, among them Bruno Latour, Jean-Louis Lyotard (who has led the assault on science), the historian Michel Foucault, and the philosopher Jacques Derrida. Their ideas spread to Britain, Germany, and especially to the United States.

One of the postmodernists' main contentions is that science wrongly claims to describe the physical world that surrounds us objectively and truthfully. In this they echo the basic tenet of relativists that all points of view are equally valid. Scientific truth, it is asserted, is only one of many truths, or rather just 'one story among many'. The American philosopher Richard Rorty, a leading and even eloquent exponent of the view that there is no truth, but only truths, ridicules those who seek the truth as 'lovably old-fashioned prigs'.[2] He maintains, with a refreshing directness unusual among postmodernists (who rarely make simple and direct statements), that truth is what your contemporaries allow you to get way with. According to postmodernists, scientists have not 'discovered' laws of nature, but have 'constructed' them. (This does not stop postmodernists travelling to international conferences by aeroplanes, whose safety, one would have thought, might be regarded as highly uncertain if the laws of aerodynamics were mere social constructs.)

Postmodernists also argue that science is wrong in claiming that the results of research can be independent of local cultural constraints or of moral and ideological motivations. In this, again, they reflect the views of relativists that all knowledge and values are relative to some particular standpoint, such as the individual, their culture and era, and so on. The work scientists do, according to postmodernists, and the hypotheses they advance are determined by where they were born, their sex, the society and culture in which they were brought up, the class they belong to, and their political ideology. Their work has to be judged by their motives and values. This view is echoed by many sociologists today and has almost become part of conventional wisdom. It was a basic assumption behind an influential House of Lords report on Science and Society and is much favoured by the think tank Demos, closely associated with New Labour.[3] It is perhaps the principal legacy of the postmodernist assault on science.

The attack on the objectivity of science is music to the ears of eco-fundamentalists. If postmodernists are right, then environmentalists do not have to worry about evidence, only about ethics. As long as they are trying to save the world, whatever they do is justified. What is more, since motive is what matters, GM technology can obviously be dismissed without regard for evidence, because it is promoted by multinational companies who, they argue, are only interested in profits.

Postmodernists also maintain, as Green fundamentalists do, that the Enlightenment, far from being 'one of the best and most hopeful episodes in the history of mankind', was in fact the precursor and generator of colonialism and oppression, and that it spread false ideas about the inevitability of progress. Furthermore, a number of modern sociologists, who propagate the view that science is not value-free, denounce it for being elitist and out of touch. It is argued that 'the public' should be more involved in almost every aspect of scientific activity to prevent its elitist bias, to allow the innate wisdom of the public to play a greater role than science has traditionally allowed and to ensure that science serves the public interest.

Science as 'the purveyor of certainties'

There were two good reasons for the rise of postmodernist influence. One was a rejection of 'Scientism', the excessive and false claims made that science deals in certainties. The other was the rejection of belief in the inevitability of progress. Isaiah Berlin, himself a champion of the Enlightenment, upbraided those *philosophes* who believed that there are universal truths which apply to matters of conduct and that we should seek a Utopia, 'a goal for which no sacrifice should be too great'.[4] Some of the *philosophes* did indeed claim that they could develop a science of human nature and answer ethical and political questions with the same certainty as those of mathematics or astronomy.

Karl Marx followed in their footsteps. He argued that the scientific method could be applied to society as well as to the world about us. In fact, he saw himself as the Newton or Darwin of the social sciences, and claimed he had discovered the scientific laws that govern human societies. Just as Newton had discovered the natural laws that determine the motion of matter in space (as was generally supposed at the time), which enable us to predict the time of sunrise and sunset accurately, so he, Marx, thought he had discovered the laws of capitalist production that enabled him to predict with certainty how societies would develop. It is said that he offered to dedicate the second volume of *Das Kapital* to Darwin because 'he had a greater admiration for [him] than for any of his other contemporaries'.[5] (Darwin wisely declined, it is reported, explaining that unhappily he was ignorant of economic science.)

Marx was wrong. Firstly, I am convinced by Karl Popper's arguments that there are no laws of history or society, because history does not have a coherent pattern.[6] Secondly, Marx's main predictions have been contradicted by events and his theory invalidated. There was no 'immiseration', or ever-growing misery of the working class. Contrary to the laws of capitalist production that Marx claimed to have discovered, it was not capitalist societies but rural economies such as Russia and China that turned Communist. Far from taking over the world, Communism collapsed. These failures helped create scepticism about science, because a central part of Marxism's appeal was its claim to be scientific.[7] Marxists believed that they understood the laws of history, that history was on their side and, because they knew they were right, they suppressed dissent.

We should not underestimate the effect that the collapse of Communism and the failure of Marxism has had on attitudes to science. Many people, including many on the left, failed to understand that its approach was anything but scientific. Communists never believed that hypotheses should be adapted in the light of criticism or experience. In the Soviet Union, anyone who challenged the theory of its leading agricultural scientist, Trofim Lysenko, that acquired characteristics could be inherited, was likely to be shot or sent to a Gulag. Yet no less an intellectual than Václav Havel, the former President of the Czech Republic, wrote in 1992: 'The fall of Communism can be regarded as a sign that modern thought—based on the premise that the world is objectively knowable, and that the knowledge so obtained can be absolutely generalized—has come to a final crisis'.[8] Fortunately, the premise that the world is objectively knowable was not a special discovery of Communism and survives its demise. Furthermore, there is no sign that disillusionment with science is today as prevalent in former communist countries as it is in many parts of the old capitalist West.

Relativism and the corruption of language

Apart from the false claims of scientism, there was also a reaction among anthropologists against colleagues who proclaimed the superiority of their own culture over the other cultures they studied. This commendable display of humility led less defensibly to the inference that facts are only true in relation to a particular culture, in other words that there is no such thing as objective truth, because no one culture is superior to another.

There are times when I have some difficulty in understanding what relativists, particularly the postmodernists, are trying to say—and this is neither a boast nor a confession—because they do not care about clarity of expression, which is regarded as a bourgeois vice and an instrument of oppression. To them, the Enlightenment is clearly the source of the ills of the world. When science was born, they argue, it provided the West with tools that enabled European nations to enslave the rest of the world—

although in fact it was during the Enlightenment that Western thinkers first proclaimed the equality of all men and their equal right to freedom and self-development. (Both John Locke and Adam Smith not only opposed slavery, but colonialism as well.) However, in the view of postmodernists, not only the applications of science and of the new technologies generated by the Enlightenment but reason itself and the language of reason aided the process of colonialization. If colonialists expressed themselves clearly, it was the duty, it seems, of anti-colonialists to express themselves obscurely. The philosopher Ernest Gellner has described the attitude of those anthropologists, for example, who swallowed the new doctrines hook, line, and sinker:

As for style ... why those colonialists wrote with limpid clarity, because they dominated the world, partly by using that wicked clarity to do so. Lucid prose and the domination went hand in hand. 'We'll show them through our style just how anti-colonialist (and pro-feminist, for that matter) we are!' And by God, they do.[9]

The style of postmodernists, as well as their rejection of the empirical approach, offered a complete antithesis to the Anglo-Saxon tradition of philosophy. As a student, I was nurtured on the philosophy of Locke and Hume, and of twentieth-century philosophers such as Bertrand Russell and Gilbert Ryle, the author of *The concept of mind*, whose writings were notable for the clarity of their prose. They are a pleasure to read. By contrast, while I admired Sartre's novels, I found *L'Être et le néant*, an existentialist forerunner of the postmodernists, almost incomprehensible. I tried to read Heidegger, often described as one of the earliest postmodernists, but the task defeated me. Mark Twain described his first editor as 'a felicitous skirmisher with a pen, and a man who could say happy things in a neat and crisp way.' No one could say that about Heidegger, even though some of my philosopher friends assure me he was a great philosopher.[10]

Heidegger's obscurity was no exception. The American philosopher John Searle relates of Derrida:

Michel Foucault once characterized Derrida's prose style to me as '*obscurantisme terroriste*'. The text is written so obscurely that you can't figure out exactly what the thesis is (hence *obscurantisme*) and then when one criticizes this, the author says, '*Vous m'avez mal compris. Vous êtes idiote.*' (Hence *terroriste*).[11]

It is tempting to draw the conclusion that the writings of postmodernists are incomprehensible because they have nothing intelligible to say.

The Sokal hoax

In fact, postmodernists sometimes cannot even understand each other. A professor of physics at New York University, Alan Sokal, was so exasperated by their confused thinking, misuse of scientific concepts and obscurity of language that he submitted a hoax account of scientific activity, phrased in suitable jargon, to a leading postmodernist journal, *Social Text*.[12] His paper, which was entitled 'Transgressing the Boundaries: Towards a transformative Hermeneutics of Quantum Gravity', was brimming with absurd-ities, full of elementary scientific howlers and non-sequiturs, but that did not prevent its acceptance for publication. This is not surprising, because its contents are indistinguishable from other postmodernist writing.

To quote from the article:

... most recently, feminist and poststructuralists critiques have demystified the substantive content of mainstream Western scientific practice, revealing the ideology of domination concealed behind the façade of 'objectivity'. It has thus become increasingly apparent that physical 'reality', no less than social 'reality', is at bottom a social and linguistic construct; that scientific 'knowledge', far from being objective, reflects and encodes the dominant ideologies and power relations of the culture that produced it ...

Again:

Liberatory postmodern science ... liberate(s) human beings from the tyranny of 'absolute truth' and 'objective reality' ... postmodern science provides a power-ful refutation of the authoritarianism and the elitism inherent in traditional science, as well as an empirical basis for a democratic approach to scientific work ... how can a self-perpetuating secular priesthood of credentiated 'scientists' purport to maintain a monopoly on the production of scientific knowledge?

So it goes.

The objectivity of truth

The first contention of relativists, that there is no objective truth, is so contrary to common sense that it is hard to take it seriously. It is not, however, unknown for philosophers to reach conclusions that conflict with common sense. When Zeno of Elea (not to be confused with Zeno the Stoic) explained to his companion that it was logically impossible to walk, his friend's answer was to get up and walk (*Solvitur ambulando*). In Tom Stoppard's play *Jumpers*, the philosopher George Moore—namesake of the famous ethical philosopher—parodies similar arguments that it is impossible for an arrow ever to reach its target (another of Zeno's paradoxes). Indeed, he concludes that an arrow could not move at all and 'St Sebastian died of fright'.[13] When Bishop Berkeley's idealism, which claimed that we could not be certain of the independent existence of material objects, was explained to Dr Johnson, he got up and kicked a stone. I am tempted to follow Dr Johnson and declare myself a member of a common sense party and to say there is no point in such nonsense. (Being suspicious of such nonsense is not the same as being suspicious of science because its findings often contradict popular, 'common sense' beliefs.)

However, dependence on objectivity and truth is the inner citadel of science. If it were overthrown, science would be pointless. How would we distinguish science from ideology, fraud, and nonsense? Fortunately, the fallacy of this basic tenet of all relativists seems self-evident. They hold that there is no objective truth. Why should we take any notice of that proposition? Is the proposition itself objectively true or not? If true, then relativists admit that propositions *can* be objectively true and their thesis is false; if it is not true, because there is no such thing as objective truth, it is a meaningless statement of which we should take no more notice than of statements that life is just a bowl of cherries or that the moon is made of green cheese. (The same objection can also be made to Marxism's claim to objectivity. If all theories are products of their class background, how can Marxism be objectively true?) Some have likened the relativist dilemma to the old paradox of Epimenides, who stated that all Cretans were liars, because he had been told this by a poet who was himself a Cretan. The comparison is inappropriate—the poet's statement was not worthless, because liars do not tell lies all the time and this may have been one time when a Cretan was telling the truth. The statement was therefore verifiable and not meaningless. Epimenides was more logical than the relativists.

The values of scientists

Perhaps the contention that should be taken most seriously is the view that science is influenced by the values and prejudices of scientists themselves, since this is a view that has become widely accepted outside the ranks of relativists and postmodernists. For example, Mary Douglas and Aaron Wildavsky argue that it is impossible to expect scientists or experts to provide an objective assessment of risk: 'Everyone, expert and layman alike is biased. No one has a social theory above the battle ... judgments of risk and safety must be selected as much on the basis of what is valued as on the basis of what is known'.[14] They too seem to accept that science can never be value-free. Likewise, the House of Lords Select Committee set up to consider the problems of Science and Society in 2000 concluded that:

Science is conducted and applied by individuals; as individuals and as a collection of professions, scientists must have morality and values, and must be allowed and indeed expected to apply them to their work and its applications. By declaring openly the values that underpin their work, and by engaging with the values and attitudes of the public, they are far more likely to command public support.[15]

The same message, that what matters most are 'the values, visions and vested interests that motivate scientific endeavour', is propagated by the New Labour think tank, Demos, in a paper that reflects the fashionable drive to make science more socially accountable.[16]

But this is plainly wrong. Both the House of Lords Committee and Demos are guilty of the fallacy that denies the objectivity of science. Of course scientists have moral and social values, but science does not, so that ultimately the motives of researchers are unimportant. Scientists may embark on a particular research project because they hope it will help mankind, or make them famous, or will confirm their prejudices, or they may select it because they can get it funded. If they work for a company, no doubt they hope it will help the company make higher profits. Whatever their motives or their values, in the end the results of their research will be subjected to objective scrutiny. Do the findings stand up to the critical analysis of peer review? Are they reproducible? Can they be verified or falsified? If the results are obviously biased by the researcher's prejudices or vested interests, they will be worthless and his or her reputation will suffer. Scientists thus have a strong incentive not to let their prejudices interfere with their work. Their reputation depends on getting things right. All scientists care about accuracy because, whatever their values, it is vital to their work. Indeed, one irrefutable answer to the supposed relevance of a scientist's background, values, and motivation is to ask the question put by Professor Robin Fox of Rutgers University, USA: What did it matter whether Gregor Mendel was a male, white, European, monk? His findings about the heritable characteristics of peas would have been no less valid if he had been a black, handicapped, Spanish-speaking, Lesbian atheist.[17]

Science is concerned with objective truth in a way that other intellectual activities are not. If a hypothesis is substantiated, it is valid at any time, anywhere, whoever thought of it, at least until a better one is found. Scientific truth is the clearest example of the philosopher Bernard Williams' description of the concept of truth: 'The concept of truth itself, that is to say, the quite basic role that truth plays in relation to language, meaning and belief—is not culturally various, but always everywhere the same.'[18] Scientific truth does not resemble the textual interpretation of a poem or of a novel, which must remain a matter of subjective opinion, about which different critics will take different views that depend among other things on their values and the age and culture in which they live. Scientific truths do not resemble political beliefs, which are inevitably influenced by our culture. Of course, there are fashions of thought, and of course, we are creatures of the age in which we were born. If I had been born in Medina in the centre of Islamic civilization in the ninth century, it is inconceivable that I would look at the world in the way I do now as a contemporary, European, liberal democrat. But if I had been an astronomer at that time in that place and had discovered a new planet, or a new law about the way the planets moved, my discoveries would have been true (or later proved false) irrespective of time and place.

Newton's work provides a good example of the irrelevance of prejudice or values or of cultural background to a scientist's work. He believed in mysticism and alchemy, devoting much of his time to the latter and, living when he did, it is not surprising that he did so. These beliefs made no difference to the validity of his scientific discoveries. I do not deny that the attitudes of scientists towards issues of scientific controversy or the direction of their work may be influenced by contemporary or personal values. When a hypothesis is still uncertain, attitudes may well reflect prejudices. When it was first suggested that smoking causes cancer, those who smoked themselves may well have been more sceptical about the findings than those who hated smoking. In time, however, the hypothesis was confirmed and today it is no longer disputed by any reputable scientist, whether they smoke or not.

Should scientists 'openly declare the values which underpin their work' in order to win public confidence, as a House of Lords Committee suggested?[19] It is not clear how this would help. Should scientists who announce a finding that the impact of pollen from *Bt* corn on Monarch butterflies in the field is negligible, declare that they have been lifelong Republicans, or Seventh Day Adventists, or disapprove of sex outside marriage? Would the public really trust scientists more if it knew what they felt about truth and beauty, or that they hate capitalism, or are Arsenal supporters? Would Newton have achieved greater public understanding of his work if he had declared his interest in mysticism and alchemy?

What seems to lie behind the suggestion is the fashionable view, part of the legacy of postmodernism, that science is not truly objective but is as much influenced by political and social values (or commercial motives) as are politicians and social scientists. Of course, decisions on the *use* of scientific discoveries and their applications may raise moral issues, for scientists as much as for anyone else. That is why Robert Oppenheimer, one of the inventors of the atomic bomb, refused to do more work on the weapon. But it is important to distinguish the results of the research itself, which is value neutral and should be judged on its merits, irrespective of the social background or motivation of those who do it, from the use to which it is put, which may raise moral problems. Almost any discovery can be abused.

The impact of postmodernism

It is easy to underestimate the subversive influence of postmodernist views on academic integrity. Some of its absurdities are more extreme than any cited in Sokal's brilliant hoax. For example, one prominent postmodernist, David Bloor, argued that Boyle's law was influenced by his conservative political beliefs and his desire to maintain the status quo in order to protect his vast Irish land holdings.[20] Feminists have called for feminist science and a feminist epistemology, to replace 'phallocentric knowledges' (whatever these may be).[21] To prove Rorty's claim that truth is anything your contemporaries let you get away with, Afrocentric historians have argued that Greek culture was stolen from Africa. One book widely used in Afro-Caribbean studies claimed that Aristotle stole his philosophy from the library at Alexandria and studied the Egyptian mystery system with Egyptian priests.[22] Therefore our culture, supposedly derived from the ancient

Greeks, originated with black Africans. Those who have tried to point out that (a) Aristotle did not visit Egypt; (b) the library was not built until long after his death; (c) Egyptian mythology and religion were quite different from that of the Greeks; and (d), apart from anything else, the Egyptians did not regard themselves as black, have been dismissed as white racists or Euro-centrists, clearly motivated by a desire to discredit the achievements of African civilization. Some American universities also teach that Socrates, because of his snub nose, was clearly black.[23]

Another absurdity is the assault mounted by some members of the constructivist-relativist school on 'the standard model of science' for being far too restrictive in its view of what is scientific. They have called for 'a reappraisal of the scientific method' to include astrology, parapsychology, psychoanalysis and other 'extraordinary sciences'.[24] After all, if science is merely one of many 'narratives', or as another leading light of postmodernism, Paul Feyerabend, called it, only one 'particular superstition', why prefer astronomy to astrology? Far more people pay attention to astrology. The astrologer on the *Daily Mail* newspaper is one of the most highly paid journalists in Britain. Another illustration of the influence of postmodernist, anti-science views on high intellectual circles is that of an exhibit in 1994 'Science in American Life' in the Smithsonian Museum in Washington DC, over which a five-year battle was fought between curators and the scientific advisory committee. The advisory committee wanted a theme of 'better living through chemistry'. The curators, who made clear their disdain for big science, wanted to expose its hazards and to show chemical manufacturers as polluters. The result was a largely negative exhibition and the waste of a great opportunity to educate the public about the excitement of science.[25]

The call for more democratic science

Any call for more democratic control is likely to be popular. If you ask people if they would like more say over almost any public issue, of course they will say Yes. Only the most old-fashioned elitist, it seems, can resist demands for more democratic control over science. But on closer examination many of such demands have no rational basis. They take different forms:

(i) The charge that science is elitist and out of touch

The Science and Technology Committee of the House of Commons set up an inquiry in 2002 into the allegation, which it clearly felt to have substance, that the Royal Society was too elitist and out of touch.[26] The House of Lords Committee previously referred to concluded that we should no longer talk about the lack of public understanding of science, because this betrayed 'a condescending assumption that the many difficulties in the relationship between science and society are due entirely to ignorance and misunderstanding on the part of the public'. There was a need for scientists to see themselves as 'civic scientists', concerned not just with intriguing intellectual questions, but also with using science to help address societal needs'.[27] At the heart of the allegation lies the belief that public misunderstandings about science are the fault of scientists, who are arrogant and fail to communicate effectively with the public.

There is no doubt that there is a mood of suspicion towards many new scientific developments and a widespread feeling that scientists should be more responsive to misgivings felt by the public. It might be summarized as follows:

New technologies and new developments in science—genetic modification and genetics are two of the latest examples and perhaps nanotechnology will be the next—affect and will affect our lives profoundly, yet in practice we have no say over what is foisted on us. Scientists work in company laboratories, or in university laboratories financed by corporate funds, and are not concerned with what we think or what we want. Because of the needs of commercial secrecy, most of the time we do not even know what they will spring on us next. If we do not have more control over the way science is going, democracy will become a sham.

Many scientists themselves now concede that science must be more responsive to the public. For instance, in 2002 the Royal

Society, representing the cream of scientists in Britain, launched an expensive campaign for meetings and discussions with the public on the theme of science and society. In fact meetings held in different parts of the country attracted few ordinary members of the public and were mainly attended by activists and cranks. The general public might express an opinion in favour of more consultation in principle, but in practice it showed little interest. It is doubtful if the meetings served any useful purpose.

I believe that blaming scientists for the current mood of suspicion towards science is to misinterpret their role. Good scientists are good scientists if they do good science. Einstein was no worse a scientist because he did not speak like Demosthenes or write like Jane Austen. When scientists are also good communicators, like Richard Dawkins or the late Peter Medawar, they can make a hugely important contribution to public education, but there are excellent journalists who inform us about science without being first-class practising scientists themselves. It is important that scientists should be open about their work and willing to explain it, but the public depends for its understanding of science primarily on the media and if there is a lack of public understanding, it is the way in which the media report science that is largely to blame. The popular belief that scientific experts misled the public about BSE and that multinational companies foisted GM foods on them without consultation are frequently cited as examples of how scientific issues have been mishandled. No doubt, with hindsight, both could have been handled differently; but BSE was a most exceptional, largely unforeseeable accident, and it is never mentioned that the public enthusiastically bought genetically modified tomato puree, explicitly labelled as GM, until the press raised the scare about Frankenfoods, based on Pusztai's now discredited experiments, a scare which was brilliantly exploited by the anti-GM NGOs.

Not surprisingly, constant demands that scientists should communicate better with the public and that they should be more socially responsible, less elitist and out of touch have driven scientists onto the defensive. One example of this defensive attitude was the presentation of a report published by the Royal Society in 2002, updating its assessment of the effect of genetically modified plants on human health. The update in fact confirmed findings of previous reports by the Society that there is no evidence of danger to health from GM food. It went on to say, very reasonably, that special care should be taken to monitor any risk both from new GM and conventional food products for allergenic effects.[28] However, the press release accompanying the report deliberately set out to suggest that the Royal Society was not making the case for GM technology and did not regard it as free from risk. Not surprisingly, journalists who did not read the report but relied on the press release duly reported that the Royal Society had changed its mind about GM food. The impression left was that support for the technology was not something to which any decent, respectable person could publicly confess or, as one commentator put it, 'it would appear that the Royal Society has not become more hesitant about the safety of GM crops and food—just more hesitant about saying so'.[29] The Society, it appears, hoped it would restore trust in scientists by hinting that it understood (and to some extent even shared) public fears.

The same reasoning, a desire to reassure the public that it was aware of its concerns, led the government to set up an inquiry into the risks that microwaves from mobile phones might be dangerous to health, despite the absence of any evidence of risk (see Chapter 7, p. 178). Far from restoring public confidence, treating unfounded fears seriously is more likely to confirm public apprehensions that the risks are real.

(ii) 'The public should have more say over new scientific developments'

The role of consultation

Clearly the public and its representatives have an important role to play in the development of science. The main lines of science policy funded by government, like other aspects of government policy, must be decided democratically by the elected government itself. Since there will always be limited funds, government must decide priorities and many policy decisions might well benefit from wider discussion. Should we, for example, spend as much public money as we do on CERN, the European laboratory for particle physics, the largest particle physics laboratory in the world, or divert some of those resources to less visionary but more practical research into plant science, where public investment is declining? Scientific developments that raise profound ethical issues, such as human cloning, issues that raise the spectre of eugenics, or the possibility, if it ever materialized, that nano-technology might enable human consciousness to be separated from the body, cannot be left to decision by a scientific elite and forced upon the public.

A report by the Royal Society on Nanoscience and nanotech-nologies published in July 2004 makes a number of sensible recommendations for public involvement at an early stage of the development of the science.[30] Effective public consultation, which may take various forms, can both improve the quality of official decisions and educate the public. One of the reasons why Britain has an enlightened system that permits research on stem cells using human embryos, is that well-informed Parliamentary debates were preceded over a period of years by public inquiries and open public discussions, reasonably well-reported in the press, at which contending viewpoints were presented in a non-adversarial atmosphere. The votes in Parliament in 2001, which resulted in large majorities in both Houses for permitting such research, were left to the individual conscience of Parliamentar-ians and were not forced through by Party Whips demanding compliance with party policy. When citizens' juries made up of representative lay citizens (not, please note, representatives from anti-science NGOs claiming to speak on behalf of the public) hear argument and evidence on both sides of a controversial issue, (e.g. how to dispose of nuclear waste or whether we should grow GM crops) without media distortion and without strident partisan advocacy by either side, they reach sensible and balanced conclu-sions. Consulting the public sensibly can often make controversial proposals acceptable. For example, experience has shown that proposals to build waste incineration plants will be bitterly opposed by local communities if public consultation is seen as a formal process for gaining public approval of predetermined plans. When there is open discussion with local residents, whose worries are seriously considered before proposals are finalized, the chances of public acceptance are greatly improved. Consultation will not win trust if it is seen merely as a way of authority informing the public of the facts.

Secrecy is also the enemy of good government. Transparency is not always possible, for instance in the intelligence services, in diplomacy, in medical research that involves the use of animals, since animal rights extremists can endanger the lives of researchers and their families, or in Budget preparations when premature disclosure of plans could disturb financial markets. But generally, the more open the processes of government, the more likely that they will command trust. The secrecy in which the Ministry of Agriculture enveloped the expert advice it received at the time of the BSE outbreak in the early 1990s only served to foment public suspicions of cover-up and conspiracy. By contrast, the new Food Standards Agency set up in 2000, which meets in public, promotes transparency, yet is dedicated to the evidence-based approach, is gradually gaining public confidence in its judgments about food safety.[31]

The limits of public involvement

Greater involvement of the public in science can therefore bring advantages. However, there are also serious drawbacks: there are limits to the useful involvement of lay opinion, there are harmful aspects to populist control of scientific research, and there is a tendency to equate democracy with the prevalence of the will of the majority.

(a) The role of lay opinion
In the previous chapter I questioned the contribution lay opinion,

as advocated by some sociologists, might make to such discoveries as the hole in the ozone layer. Currently, demands for more account to be taken of lay opinion are commonplace. The Phillips report on BSE suggested that we should take note of the views of victims' families.[32] Reports on railway crashes recommend that we should take note of the views on policies for railway safety of those injured or the families of those who were killed. The Stewart report on mobile phones recommended that in future 'non peer-reviewed papers and anecdotal evidence should be taken into account'. In the media coverage of the MMR controversy, the opinions of parents of autistic children were given equal weight to those of scientific experts on MMR and autism.

Of course we should listen to the experiences of victims and their relatives. But victims of railway crashes, however heartbreaking their experience was, do not automatically become experts on how to run a railway. As a result of treating them as experts, a safety system was recommended for the railways in Britain that would cost £3 billion and save up to five lives a year, when ten people a day are killed on the roads.[33] Nor do victims of disease or their relatives automatically become medical experts or experts in healthcare management. People do not become authorities because they are in the news but because they are reliable and reliability does not depend on having their heart in the right place or being well-known but on acquiring opinions by a reliable method, that is with knowledge of, and regard for, evidence. Regarding lay opinion or anecdotal evidence as equivalent to peer-reviewed scientific findings is another legacy of the post-modernist view that science is 'just another story' with no special claim to objectivity.

Again, some sociologists want to see a greater lay input into the assessment of risk, though risk assessment requires expert statistical analysis. (How to deal with public *reaction* to risk requires a very different expertise, including the kind of understanding that lay people may be better able to provide than scientific experts.) This was not, admittedly, the view of the House of Lords committee on Science and Society, which claimed that 'the public understands risk well, on the basis of everyday experience'.[34] Statistical analysis is not, however, one of the public's strongest suits. Less surprisingly, Green activists, who reject 'the probabilistic, rational approach', also strongly support the view that lay opinion must play a central role in risk assessment.

Clearly, the call for more public involvement cannot be justified by the argument that public opinion must be right by definition and that to assert otherwise is undemocratic. Many campaigning organizations (including consumer groups) frequently cite public belief that GM crops or pesticide residues in our food are unsafe as conclusive evidence. But what basis can there be for the belief that the public has an instinctive grasp of what are mainly technical issues?

Take, for example, decisions about the level of pesticide residues in food that can be considered safe. Deciding whether the concentration of a particular chemical is harmful is a technical process that depends entirely on expert knowledge. First, the toxicological profile of the chemical must be established, hazards must be identified and characterized, then the presence of that compound must be detected in the food in question, then the amount present must be measured, then it must be determined whether this amount will cause harm to human health. The last stage is the most complex and expert opinions will be exchanged, but on the basis of the best available evidence a committee of experts will set the level of acceptable daily intake, which will be many times above the perceived minimum safe level. This normally allows a margin of safety of approximately 100–1000 times for consumers. The experts would all be blamed if they were not ultra-cautious and someone was to be poisoned as a result.

At least this is the rational way of proceeding, but it is not always applied. For instance, when the European Union sought to replace the precautionary limit for total pesticide residues in drinking water by a science-based standard, it abandoned its proposal as 'politically unsaleable' after receiving 12,000 protest letters in a campaign organized by Greenpeace. Regulating risk by listening to activists and complainers has been likened to the 'scream' tradition in government budgeting, with allocation of funds based on the volume of screams.[35]

How can popular instinct, or the special insights of activists, improve on the assessment by experts in deciding safe levels of pesticide residues? Can the public, or the activist, detect the presence of the chemical by osmosis and know how much of it is harmful to health by divine revelation? How can intuition get it right, except by pure accident? If consulted in calm, unemotional circumstances, ordinary people will generally acknowledge that the assessment of risk that depends on technical knowledge should be left to experts. Unless you are a Christian Scientist, a Jehovah's witness, or so addicted to alternative medicine that you reject conventional medicine altogether, you expect the diagnosis as to whether you have a brain tumour to be carried out by a specialist. If you then need an operation, you ask a brain surgeon to operate.

(b) Public control of science

Ten years ago, a consultancy called SustainAbility proposed that companies should carry out a needs test before they proposed a new product or service. The Demos pamphlet 'See-through science' strongly approves. Throughout it refers to public concerns (which it clearly regards as legitimate) that a technology must be 'needed'. Who will control it, they ask, who will benefit? To what ends will it be directed? It is also argued that there should be no more 'science for science's sake'.[36] In effect it wants all new developments to be under popular control.

Such views show little understanding of the history of science or indeed of what science is. Any number of invaluable scientific discoveries and technological inventions, from Faraday's discovery of electricity to the invention of the laser, came to have uses that no one foresaw, or could have foreseen. Should they have been rejected because no 'need' was proved? The call for more democratic control of science soon becomes a demand for political control over research. 'The people', it seems, are to decide what is allowed, because they, not scientists themselves, must determine the purposes for which scientific developments can be used.

Furthermore, scientific research will only be permitted if it is strictly directed at utilitarian ends which 'the people' approve. No science for the sake of science. Scientists and science are clearly too dangerous to be allowed to pursue knowledge for the sake of better understanding of the world around us. Those who favour a needs or utility test seem unaware that to seek knowledge for its own sake is one of the noblest endeavours of mankind.

But why should science be singled out as needing more democratic control when other activities, which could be regarded as equally 'elitist' and dependent on special expertise, are left alone? Why not more democratic control of sport or the arts? Is there also to be a ban on 'Art for Art's sake'? Science, as well as the arts, depends on inspiration and such creative geniuses as Galileo, Newton, and Einstein can fairly be compared with Michelangelo, Shakespeare, and Beethoven.

The scientific process, has never been a democratic one. It is not a search for some convenient consensus based on compromise; indeed consensus is irrelevant in science. What matters is not how many people declare their support for a theory, but whether a hypothesis stands up to critical analysis and whether experiments are reproducible. Nor is the purpose of science, like the aim of governments, to make us feel good.[37] Furthermore, many of the conclusions of science contradict popular beliefs and are as likely to disturb as to reassure. Science is concerned with our understanding of the world, a search for truth that does not depend on whether the public believes it or likes it. Had there been a referendum in Galileo's time as to whether the earth went round the sun or vice versa, public opinion would almost certainly have rejected his view and supported the Inquisition.

Popular, that is political, control over science, over the research that scientists want to pursue or the publication of its results, has always proved fatal to good science, just as state control is the death of art. The scientific revolution of the Enlightenment became possible because the domination of the church was broken. What was true and what people might be allowed to think could no longer be decided by reference to the bible. Galileo's magisterial protest that the authority of the almighty church should not interfere with the truth-seeking activities of science and his ridicule of the idea that an 'absolute despot, being neither a physician nor an architect, but knowing himself free to command, should undertake to administer medicines and erect buildings according to his whim' is just as valid if the demos (or the think tank Demos) were substituted for the despot. Since pre-Enlightenment days there have been periodic attempts to re-impose ideology and the results were invariably disastrous. Lysenko's ban on 'bourgeois' genetics, for example, held back the adoption of new hybrid seeds that had been developed for Western agriculture, and Soviet science was the loser. German science took a long time to recover from Nazi views that orthodox science was Jewish science and Jews should not be allowed to practise it.

Those who want more democratic control of science will protest that control by democrats is by definition the opposite of control by communists or fascists. However, the demand for more public input into the subject matter of research, whether in the name of democracy or ideology, would still suppress the independence of science. (The demand for more public *finance* of research, as argued earlier, is eminently sensible, just as public support for the arts need not imply public control of what an artist produces.) More public input would in practice mean public supervision to counteract the 'elitist' character of science and make it more responsive to public consensus. This would mean control by committee and the first result would be a bias against excellence and unorthodoxy. Committees and genius do not mix well.

A second result would be insistence on political correctness. After all, one reason for demanding democratic control is the belief that science itself cannot be value-free. Committees would ensure that the values of research are the right values, values that are politically correct. When Green NGOs refer to public opinion, they mean activist opinion. The British Government's choice of lay appointees to various advisory bodies on scientific policy, such as the Agriculture, Environment and Biotechnology Council, has made the same mistake: representatives from the same NGOs—Greenpeace, Friends of the Earth, Genewatch (an off-shoot of Greenpeace), and the Soil Association—appear and reappear like the same stage army in battle scenes of Shakespeare's histories. The Green pressure groups make the most noise and are the best organized. It is most likely that their view of political correctness will be imposed.

The warning signs are clear from American experience, where in deference to public opinion attempts were made to suppress publication and to ban conferences on highly sensitive subjects such as the correlation of criminal behaviour or sexual orientation with a particular gene. Why allow research that may have dangerous social consequences?[38] Research into racial characteristics has always been regarded as particularly dangerous, whereas in fact its results are more likely to destroy than promote racist misconceptions. Even non-controversial assumptions about race are almost invariably unfounded. Those who believe, for example, that black basketball players have greater natural aptitude for basketball than whites, forget that in the 1920s and 1930s basketball was dominated by Jewish players. It was generally thought at the time that they had the advantage that their race gave them sharper eyesight and a special aptitude for quick movement. At about the same period, nearly all American jockeys were black.[39] Political correctness, whatever its more general justification, conflicts with free speech and, in science, can only inhibit the search for truth.

(c) Misconceptions of democracy

Fundamentally, the demand for more democracy in science and its antagonism to 'elitist' science reflects a belief that the people's will must prevail and that expert minorities must serve the interests of the majority. In general that proposition holds good. However, democracy is not simply a question of ensuring the triumph of the people's will. It requires a balance between the wishes of the majority and the rights of minorities, between respect for expertise and the need to ensure that experts serve the interests of the majority. It is not self-evident that in a liberal democracy the will of the majority should always be decisive.

Take two controversial examples that involve both ethical and evidential considerations: firstly, should the question as to who should receive public health care when funds are limited be decided by the legislature or by more direct consultation of the public? Secondly, should the issue of capital punishment be left to parliaments or decided by referendum, to ensure that the views of the majority on how to deal with the most serious of crimes are properly respected?

In the first case, a widely-based public consultation exercise in Oregon in the USA found that there was strong public opposition to spending limited public funds on AIDS or mental health.[40] In the second, if a referendum were held in Britain, according to opinion polls, it would lead to the restoration of capital punishment. Not everyone would regard the outcome in either case as a triumph for liberal democracy. It is perfectly democratic to argue that since the protection of minority rights is an essential element of democracy, popular indifference to the sufferings of the mentally ill or the affliction of gays should be ignored. In Britain, despite public support for capital punishment, Parliament, after full argument and careful consideration of evidence, has consistently decided against its restoration. Two arguments in particular carry weight with the legislature: contrary to popular belief, there is no evidence that the death penalty reduces the number of murders, and from time to time, even with the most careful legal safeguards, innocent people will be convicted for whom no posthumous pardon or reprieve can restore justice. Of course, when Parliament flouts popular opinion, the majority may always get their revenge, since the people can always vote out of office those who ignore their views.

There are other reasons why automatic acceptance of majority opinion does not always serve the interests of democracy. Public opinion is often fickle. If laws had to reflect popular sentiment at all times some would have to be periodically repealed, reinstated, and perhaps repealed again. A Dangerous Dogs Act was passed in

Britain in 1991 after some well-publicized cases in which children were savaged by dogs. It fell into disuse when sympathy shifted to the owners of much-loved pets ordered to be put down by court decision. Laws that are unpopular when passed can become popular in practice. Thus, legislation against drink-driving and the imposition of congestion charges on cars in London were both introduced in the teeth of popular opposition but later proved popular. Sometimes, therefore, governments are justified in ignoring the popular will and giving a lead.

In fact, there are many decisions that affect people's lives profoundly that governments delegate to groups of experts. In the Euro-zone, in Britain, and in the United States crucial decisions over the future of the economy are delegated to central banks whose boards are unelected and which consist exclusively of economic or banking experts. Yet it can be argued that decisions on monetary policy are no more technical, or less amenable to lay challenge, than issues such as the setting of minimum safety levels for pesticide residues in food. Universities are elitist organizations that select entry on the basis of talent and ability to profit from further education, although most countries seek to maximize the number receiving higher education. Professors are not elected by popular vote. Judges who decide important issues of justice are not elected, at least not in Europe, but appointed for their expertise and other judicial qualities. The more 'democratic' American system of electing most judges has few admirers outside the United States. Not the least of its demerits is that it politicizes the judiciary. For instance, President Ronald Reagan's appointment in 1986 of William Rehnquist as Chief Justice of the US Supreme Court was made with the express intention, since fulfilled, of reversing two decades of legal liberalization.

It is true that we leave it to a jury of ordinary citizens to decide if their fellows are guilty of serious offences, but more complex issues of civil law are left to an expert judge. Britain abolished the role of juries in civil proceedings long ago, except in, relatively rare, libel cases. (The general view among barristers in cases where there used to be a choice between a trial by jury or by a judge sitting alone, was that if you had a bad case, you chose a jury; if you had a good case, you chose a judge.) Critics of the American system, where juries still decide issues involving complicated technical and scientific evidence in patent law and negligence claims, argue that juries in civil cases often reach decisions, both in their verdicts and the damages awarded, that can only be described as perverse. In patent cases their verdict is widely regarded as a lottery. In claims of medical or environmental negligence, sympathy for plaintiffs, or antipathy towards rich defendants, often seems to outweigh regard for evidence. In her book *Science on trial*, Marcia Angell, a former editor of the *New England Medical Journal*, has described the perverse verdicts reached by juries in cases about breast implants. Huge damages were awarded although claims for negligence were based on non-existent, scientifically unproven, links between silicon implants and connective tissue diseases such as rheumatoid arthritis.[41] The experience of American juries in civil cases is not a convincing argument for giving lay people more say over complex scientific issues.

In conclusion, science has been attacked by postmodernists and relativists because it represents reason and reason has gone out of fashion in parts of academia. The call from influential Green lobbies that science must be more democratically accountable finds a sympathetic response from the public because health scares (and some public mismanagement) have made the public suspicious of experts in general and of scientists in particular. Some forms of greater public involvement with science may reduce suspicion and improve understanding, but public distrust will not be dispelled by abandonment of the uncompromising commitment of science to the pursuit of truth or by questioning its objectivity. '*Nullius in verba*' is the motto of the Royal Society: On the Word of No One. It is one of the most important messages the Enlightenment brought to the world.

9

Multinational Companies and Globalization

Capitalism should be replaced by something nicer.
Banner of an anti-globalization demonstrator at Genoa

ONE of the reasons why opposition to GM crops is particularly strong on the left in politics is because it is almost inextricably intertwined with anti-capitalism.

Hostility to genetic modification and hostility to multinational companies go hand in hand. According to an editorial in the journal of the New Economics Foundation: 'The truth is that growing GM crops is actually about the manipulation of international patents to create guaranteed markets for monopolistic agri-businesses'.[1] It is unlikely that eco-fundamentalists would have an open mind about GM crops even if big business were not involved, but the link between the two is another reason for demonizing the technology. Earlier I cited opinion polls showing that the public mistrusts scientists who work for private companies or whose work is financed by them. It is widely assumed that the profit motive undermines the integrity of their research. Suspicion of big business has also been a factor in the reaction of parents against the MMR vaccine, because they believe that the large pharmaceutical companies who manufacture the vaccine are not interested in public health, but only in maximizing profits. Then there are aid agencies who share the belief of Green lobby groups that multinational companies do not promote products that poor Third-World farmers actually need because they only want to exploit new markets for their own benefit. Finally, international companies, agri-businesses, and pharmaceutical giants prominent among them, are denounced by demonstrators against globalization as the villains of international trade who perpetuate world poverty and increase inequalities between nations. This makes many people who feel passionately about injustice in the world lump together big business and the technology it promotes as the common enemy.

Opposition to big business has become so obsessional in some quarters that elaborate records are kept, published from time to time in sections of the media, of anyone involved in the GM debate who has ever had any connection, however remote, or received any financial support, however indirectly, from any major corporation. It is then assumed as axiomatic that their research findings or comments on scientific issues must be worthless. These extreme conspiracy theorists have categorized the Royal Society, for example, a body to which Britain's most eminent scientists are elected, as biased and untrustworthy because it has received corporate support. Any research financed by the Wellcome Trust, now the biggest source of support for medical research in Britain, has been declared as tainted because the Trust was originally funded from the profits of a pharmaceutical company.[2] By the same standard, all major charitable foundations in Britain must be suspect, because all were originally funded out of corporate profits. The Rockefeller Foundation in the United States (likewise no doubt many other American charitable trusts) is doubly damned: its source of funds was originally one of the world's biggest capitalists and it supports research into GM crops.

The obsession in Green lobby circles with the source of finance is well illustrated by references to Arpad Pusztai, whose research played a major part in originating the 'Frankenstein food' scare. Experts from the Royal Society and other authoritative bodies condemned his work as unsubstantiated and unreliable. Nevertheless, it is still frequently quoted by anti-GM campaigners as evidence of the danger of GM crops to human health, not because it has been supported by further evidence, but simply because his critics at some time had support from capitalist sources. What matters to conspiracy theorists is not the message but the messenger, not the value of any piece of research, but the supposed motives of the researcher. Of course it follows that, if all scientists who have ever been supported by corporate finance are disqualified as biased, about the only trustworthy scientists left are those who are employed or financed by Green NGOs.

It is an irony of history that since the triumph of capitalism over communism, people in the capitalist West, especially many young people, have grown more, not less, hostile to capitalism. Just when most countries accept that the market system, not socialist planning, is the way to create wealth, distrust of organizations that trade in the market and make it work, namely private companies, and of the engine that drives them, the profit motive, is as widespread as ever. Even branches of the Women's Institute, once the apotheosis of conservative, middle-class respectability in Britain, have joined demonstrators against globalization and multinational capitalism at Seattle and Genoa. Paradoxically, it seems we are all anti-capitalists now, though some more so than others.

Several factors explain the paradox. Firstly, capitalism has become the victim of its own success. Now that it has vanquished Marxism, to be against capitalism no longer carries the taint of being in league with communism. It has become respectable. Many Green parties were formerly as strongly anti-communist as anti-capitalist but now feel free to attack capitalism without any sense of inhibition.

Secondly, some capitalists have proved their own worst enemies and hubris has prevailed over morality and good sense. Self-restraint has been abandoned in an orgy of free-market triumphalism. Particularly in the United States, there appears to be a feeling that since capitalism no longer has to justify itself against its former rival, socialism, self-interest and greed can be allowed free rein. Thus, the directors of big companies pay themselves enormous salaries under the pretence, wholly uncorroborated by evidence, that they need them as incentives to be good managers. Throughout the 1990s, board members in America were given options to buy the shares of their own companies cheaply so that they would benefit directly from a higher share price. As a result, in the ten years to the start of the twenty-first century, the cumulative value of share options rose tenfold, to some US$600 billion, 'one of the greatest wealth transfers recorded in history'.[3] Senior directors are also awarded huge sums as compensation for loss of office even when they have demonstrably mismanaged the companies entrusted to their care. This happens not only in the United States. As one British observer put it, a new principle has been born: 'failure must never go unrewarded'.[4] To compound the offence, some chief executives of top American companies such as Enron and WorldCom added fraud to greed. No wonder the American model of capitalism has acquired a certain tarnished look and people feel that boards of big companies are only concerned to feather their own nests.

A third factor, and perhaps the most important, is the rise of popular support for environmentalism, now a powerful political force that has hitched its wagon to the anti-capitalist cause. Interest in conventional party politics is declining, certainly in the Anglo-Saxon world, because they seem increasingly irrelevant to people's daily lives. On the continent of Europe too, 'social capitalism' has largely abolished the traditional conflict between left and right. Only the occasional eruption by Jean-Marie Le Pen or Jörg Haider perturbs the general consensus. In Britain, New Labour, like the Democrats in America, occupies the centre ground of politics, and passion in party politics is reserved for special issues: in Britain, whether we should adopt the euro or leave the European Union; or lately, whether we should have gone to war against Iraq or whether fox hunting should be outlawed; in America most passionate debate in conventional politics (apart from the debate about Iraq) seems to centre around moral issues such as abortion. Environmentalism on the other hand is a radical movement that does arouse political passions. Saving the planet is a cause worth marching for. Many people care deeply about the loss of rain-forests, or the potential extinction of the panda or the whale, or about threats to their local environment, whether it is the building of a by-pass or plans to dump nuclear waste in their backyard. Unlike political speeches at party meetings, environmental issues gain attention in the media, especially when protests against polluters, generally big companies, can be dramatized as battles between David and Goliath. When disillusioned veterans of stale party warfare join demonstrations to preserve nature and to denounce the wickedness of multinational corporations, it can give them a nice warm feeling that the old radicalism is still alive.

Unlike the 'Back-to-Nature' movement of the first part of the last century, which was traditionally associated with the political right, today's active environmentalists are the new left. They have linked up with champions of the world's poor, NGOs, and church groups protesting against Third World debt, with former Marxists who no longer have any major party to express their views, and with anarchists and other malcontents who oppose all authority and like nothing better than a riot. Together they form a motley alliance of disparate groups, pacifists and militants, lambs mingling with lions, who focus their various discontents against one common target: globalization. Anti-globalization has replaced nuclear disarmament as the cause that stirs the blood and attracts more young people and bigger demonstrations than any political movement of the recent past (until the war in Iraq).

Globalization and the nature of modern capitalism are big, complex issues that go far beyond the scope of this book. Nevertheless, because they are so closely linked with attitudes to science, they are issues that cannot be ignored. If multinational companies, that are nowadays mainly responsible for the development of new technology, are also the driving force behind globalization, if free trade and globalization, as their critics claim, increase inequalities, cause international injustice, worsen environmental pollution, and undermine attempts to feed the hungry and raise the living standards of the poor, and more generally, if the nature of capitalism corrupts the integrity of science, then suspicion of the role of science in modern society would be justified.

I am not an apologist for big business. Many activities of many international companies, for example tobacco companies, deserve unqualified condemnation. There have been a number of incidents in the past where multinationals interfered in national politics, or persuaded their government to do so, particularly in South America.[5] In domestic affairs, there is clear evidence that the marketing departments of pharmaceutical companies have often brought improper pressures to bear on doctors to recommend their drugs and on editors of scientific journals to review the results of their company's research favourably. The increasing dependence of universities on financial support from industry can lead to abuse and has sometimes done so. In one notorious case, a research worker in the University of Toronto found that a drug developed by a company that sponsored her work, and which was planning to make a large donation to the university, was less effective than expected and had serious side-effects. The company sought to suppress her work and she was dismissed. After a long investigation she was eventually reinstated[6] and the furore her case caused will act as a powerful disincentive for similar abuses elsewhere. Another disadvantage of academic dependence on corporate finance is the rush by academics to take out patents on their work. This prevents scientists from sharing information about research results and undermines good science. Paradoxically, the excessive scope of patents now being granted is hampering the innovation for which patents were invented.

There is no doubt, therefore, that business activities need effective regulation and control, which does in fact exist in most of the developed world. On the other hand, I reject the view of the conspiracy theorists that all multinational companies sacrifice all ethical considerations for the sake of profit. Most companies believe that the profit motive is not incompatible with serving the community and that by creating wealth they are making the world a better place. Furthermore, I believe that, in general, free trade promotes human welfare, that the World Trade Organization is too weak rather than too strong and should be strongly supported, and that most of the criticisms made by the anti-globalization movement of the role of multinational companies are as wide of the mark as the allegations made by anti-GM crusaders about the dangers of GM crops. Finally, to damn all research results because they are based on corporate finance is to repeat the postmodernist fallacy of confusing the validity of a statement with the motives of the person making it and denying the objectivity of scientific truth.

The case for free trade

The basic case for free trade has not changed since the days of Adam Smith, who explained in 1776 in *The wealth of nations* that if you consume only what you can produce yourself, you will be less well off than if you specialize in producing what you can make better than others and exchange it for what they make. The same principle applies between countries as between individuals and businesses.

There is strong evidence that the world benefits greatly from free trade, and that this is true of poor countries as well as rich ones. Compare, for example, the period since the Second World War with the period between the wars. Protectionism was the rule during the 1920s and 1930s and was accompanied by mass unemployment, whereas the period of fifty years and more after 1945 was a time of free trade, which brought a long boom and steady economic growth. Developing countries that adopted free trade and opened their frontiers, such as Singapore and South Korea, lifted themselves out of poverty, while those that kept up the barriers and sought to remain self-sufficient were left far behind (compare South Korea with North Korea). Today, China, which has adopted free trade, has one of the fastest-growing economies in the world and increased its real income per head more than fourfold between 1980 and 2000. The GDP of developing countries with open economies grew by 4.5 per cent a year in the 1970s and 1980s, while that of those with closed economies with high-tariff barriers and import quotas grew by only 0.7 per cent.[7] Equally impressive is the contrast in India between the period of near stagnation and protectionism during the 1960s to the mid-1980s and the period since then when the Government started to liberalize the economy, leading to a dramatic spurt in its growth rate after 1991. As a result, real income per head more than doubled between 1980 and 2000.[8]

The last twenty years have not only seen a dramatic rise in prosperity in East and South Asia, where 55 per cent of the world's population lives, but also the fall of the Soviet empire and the spread of democracy to many parts of the world where it was previously unknown. Most of us would regard this as a cause for celebration. The anti-globalization movement, however, regards it as a period of catastrophe.[9] It is true that some parts of the world have not prospered and indeed have grown even poorer. The continent of Africa remains a region of widespread poverty, at least partly because it has hardly been touched by free trade and has attracted little investment by multinational companies. Indeed, the biggest single advance in the prosperity of some of the world's poorest countries we could make today would be to extend free trade further, to allow poor Third World farmers better access to the protected agricultural markets of Europe and North America, reduce trade-distorting forms of domestic support, and eliminate export subsidies for prosperous American and European farmers. Such radical reform would bring far greater benefits than increased international aid, even if it were increased to levels that most Western nations are unwilling to pay. The end of trade barriers between developing countries themselves would have at least equally beneficial effects. As Kofi Annan, the Secretary General of the United Nations, has often stressed, what the Seattle protesters and anti-globalization movement do not realize is that Third World countries want access to trade. They want to trade themselves out of poverty, not live on assistance.[10]

Findings from polls strongly support Kofi Annan's statement. A worldwide poll carried out by the Pew Global Attitude Survey in 2002 among 38,000 people in 44 nations found that people in low-income countries are more in favour of globalization than people in rich ones.[11] For example, in sub-Saharan Africa (where there has been relatively little international investment) 75 per cent of households thought that multinational corporations had a good influence on their country, compared with only 54 per cent in rich countries. In Uganda and Vietnam, countries that have benefited substantially from foreign investment, the proportion who thought that growing economic integration was 'very good' for their country was 56 and 64 per cent respectively; by contrast only 28 per cent of people in the United States and Western Europe thought so. Anti-globalization protesters were approved of by 35 per cent of people in rich countries, but only by 28 per cent in Africa. Protesters claim to speak for the poor of the world, but the poor do not seem to know it.

The impact of free trade on inequalities of wealth is not easily calculable. According to figures supplied by the World Bank, which on the whole takes a more pessimistic view of the decline in world poverty than other economic estimates, the total number of the very poor living on less than $1 a day declined in the 1990s from 29 to 23 per cent of the world's population. In 2002 between-country inequality was reducing (largely because of the economic growth of China and, to a lesser degree, India), but within-country inequalities (both in developing and developed countries) were in some cases increasing.[12] However, while inequalities have lessened overall, in some parts of the world absolute as well as relative poverty has increased, notably in sub-Saharan Africa. What the evidence seems to show is that where free trade is extended, poverty is reduced. Where poverty has worsened, or not changed, the main reasons have little to do with free trade, much more with war, weakness of infrastructure, corruption, and the tragedy of AIDS and other diseases. Nevertheless, the overall picture supports the case for free trade, as it shows that where free trade extends, poverty is reduced, whereas in areas relatively untouched by free trade, there is little, if any, improvement.

As the Pew worldwide poll revealed, free trade is not popular with everyone in rich countries. In fact it is mainly the rich who cry foul and complain of unfair trading. In some ways, it is surprising that politicians in the advanced industrial countries have been willing to support free trade at all, since the painful consequences are immediately evident to particular (and often very vocal) pressure groups, while its benefits are longer-term and less obvious because they are more widely spread. The immediate effect is likely to be a loss of jobs. Hence the loudest protests about low wages and about sweatshops in the developing world come not from champions of a more equitable distribution of the world's wealth but from its opponents—unions that represent relatively prosperous workers in the northern countries and that see their members' jobs endangered by competition from cheaper and poorer producers in the south. Naomi Klein, the author of *No logo*, one of the most influential books about the evils of globalization, sometimes takes a straightforwardly protectionist view. For example, she cites with obvious sympathy, complaints from two members of the US Congress that

Nike has led the way in abandoning the manufacturing workers of the United States and their families . . . Apparently Nike believes that workers in the United States are good enough to purchase your shoe products, but are no longer worthy enough to manufacture them.[13]

The same protests now denounce the 'outsourcing' of American (and British) jobs to low-paid Asian workers. What is this but protectionism pure and simple, and indefensible? When jobs are exported from the United States to the Philippines or Indonesia, of course American workers suffer and complain, but poorer countries gain. In fact, most blame for the decline in demand for low-skilled workers in the northern hemisphere is due not to loss of jobs to the developing world, but to technological change.

The case for the World Trade Organization (WTO)

No one claims that the WTO is perfect. It is a small body (not a huge bureaucracy, as those who demonstrate against it imagine) with a budget of less than a quarter of that of one of its critics, the World Wildlife Fund. It is limited in what it can achieve since its decisions depend on agreement among all of its 140-plus members. Inevitably, since its role is the regulation of international trade, big trading blocs carry a much bigger stick than small groups of nations. However, at a meeting in Doha in 2001 it was agreed to promote the so-called Doha Development Round for the benefit of developing countries. Although the next meeting of the WTO at Cancun in 2003 broke down without agreement, at the time of writing all the major groups involved have shown a determination to make renewed progress and it seems likely that future liberalization of trade will be achieved and that this will mainly benefit the poorer nations.

The WTO benefits rich and poor nations alike. Environmental lobby groups such as Greenpeace complain that the WTO should be more accountable and democratic and that this should be achieved by including representatives from organizations like themselves. Lack of accountability is not, however, the trouble with the WTO. It responds to national governments. Why should being responsible to Greenpeace, which has no democratic structure, make it more democratic than having to win the consent of governments? Are the activists of Greenpeace more representative, for instance, than the Government of India?

The WTO has more power than the organization it replaced, the General Agreement on Tariffs and Trade (GATT) because once rules of international trade have been agreed, WTO rulings are binding. Unlike those of its predecessor, they cannot be vetoed by individual states. That is why the protesters are wrong to attack the WTO: uniquely in the international scene it is a binding mechanism for the settlement of trade disputes. With the chief exception of agriculture, most protectionist measures are outlawed and countries that have suffered harm from them may impose penalties on offenders. For instance, much to the chagrin of European Governments, the WTO ruled that the EU ban on hormone-treated beef was protectionist and therefore illegal. The EU had claimed that the ban was justified to protect health because such beef might be carcinogenic. (WTO members may impose restrictions on environmental or health grounds.) However, the EU's own scientific committee had ruled that there was no danger to health. Accordingly the US was allowed to impose trade sanctions, the cost of which are likely to exceed $200 million.[14]

The WTO gives poorer countries a chance to argue their case against the rich and to win if their case is a good one. Ecuador, for example, a country with an annual income per head of about £1000 a year, was prevented by the protectionist rules of the European Union from selling its bananas in Europe. Instead it appealed to the WTO, which ruled in its favour. In due course, Ecuador will be able to sell more bananas in the EU. The WTO upheld a complaint by Costa Rica that the United States was unfairly preventing its exports of underwear and it made the mighty US lift restrictions. Mexico has also won trade disputes with the United States through the WTO. What anti-WTO protesters fail to see is that the alternative to the rule of law is the rule of the jungle, where might is right and the big beasts always win. Without the WTO weak nations would be even weaker, not stronger.

Reforming the treaty on intellectual property rights

There is one respect, however, in which the interests of developing countries have been prejudiced: in the agreement on intellectual property rights, such as patents. In 1994, as part of a wider trade deal, members of the World Trade Organization (WTO) agreed a treaty known as TRIPS (trade-related aspects of intellectual property rights) that sets out minimum standards for the legal protection of intellectual property. The treaty was to be implemented at once, with a period of grace for developing countries, which has since been extended to 2005, and to 2016 for those who are least developed. A Commission was set up by the UK Department for International Development (DFID) to consider the effect of TRIPS and its conclusion was that the overall impact was to favour the rich and prejudice the poor and that the treaty needed reform. The countries which pressed most strongly for its adoption were, not surprisingly, those that have an industry to protect: the World Bank estimated that the United States, for example, would gain some $19 billion a year from the treaty through the enhancement of its patents.[15] Developing countries, on the other hand, will in due course pay more for patents that do not apply to them at present, and generic producers, which now flourish in countries like India and Thailand, may have to close once foreign patents are given the force of law locally.

It is argued by defenders of TRIPS that developing countries can avoid many of its disadvantages and have much to gain. Each country is free to introduce its own patent laws. At the WTO meeting in Doha in 2001, it was agreed that this would leave them free to take measures to protect public health and would also allow them to manufacture or import copies of a drug without the patent-holder's approval (but subject to a payment of compensation) under a system of so-called 'compulsory licensing'. Parallel imports would be permitted, that is the importation of a patented drug from cheaper sources elsewhere. At the same time, it was argued that developing countries would gain positive benefits: some would attract more foreign investment by adopting a patent regime and thereby promote the transfer of technology; others, like India and China, could develop their own biotechnology industry, as the developed world has done, by enacting their own patent laws; and others would be better able to protect their natural plant resources and exploit their economic potential.

Most of the claims for benefits from TRIPS for the developing world ring hollow. In fact intellectual property rights are irrelevant to the needs and problems of most developing countries. India, China, Brazil, Mexico, and South Africa may in time need a domestic patent regime (although TRIPS is likely to damage industries making generic drugs), but most developing nations, especially the poorest ones, will not have any innovations and discoveries to protect. There is no clear evidence that foreign investment or technology transfer is promoted by protection for intellectual property (IP). The DFID Commission also found that IP protection played no role in stimulating research on diseases prevalent in developing countries, except for those diseases (diabetes and heart disease are prominent examples) where there is a large market in the developed world as well. There are no patents on trypanosomiasis or diarrhoeal diseases; in 94 per cent of countries surveyed by the World Health Organization there were no patents on TB and malaria drugs and no countries had patents on all the relevant drugs for these diseases.[16]

The special problems of access to the drugs they need for developing countries, especially those without their own manufacturing capacity, were recognized in the WTO Ministerial Declaration on TRIPS in 2001. No one wants to renegotiate the treaty. Instead the special Council that exists to implement TRIPS was instructed to find some solution to enable developing countries to overcome their access problems. The DFID Commission made a number of technical recommendations on how the treaty should be implemented: the scope of patents in developing countries should be severely restricted, mainly to exclude the patenting of animals and plants and to avoid using the patent system to protect plant varieties and, where possible, genetic material. However, the main problem remains access to drugs. There is no simple solution.

The high price of drugs, for example, is not the only issue. The reason why recent controversy about drug prices has been about the cost of anti-retroviral AIDS drugs, is that AIDS is one of the few diseases that afflicts the developed and developing world alike. Very few other drugs have yet been developed that meet the needs of poorer countries. Pharmaceutical companies aim to market drugs to prevent or treat diseases of aging populations in the west, such as Alzheimer's disease and arthritis, rather than malaria or diarrhoeal diseases. Significantly, two of the most profitable drugs of recent decades have been Prozac, to treat depression, and Viagra, to dispel impotence. These are not diseases that ravage sub-Saharan Africa. It has been estimated that only 13 of 1393 drugs approved between 1995 and 1999 were specifically intended for tropical diseases.[17]

Next, even if the right drugs are available at fair prices, poverty itself stops them reaching those who need them. Poor countries have no adequate means of drug storage and delivery; there are no properly trained staff to administer vaccines, no syringes and needles to inject them, and no refrigeration to store them in good condition. Furthermore, patients may not be able to follow a strict regime of regular ingestion or application (hence the enormous benefits that would flow from the delivery of vaccines through genetically modified tomato juice, bananas, or potatoes). Inadequate roads or local wars prevent drug distribution and in many countries corruption diverts funds into the wrong hands. The problems of recipient countries are so great that the WHO has estimated that any country that spends less than $60 a head on health is not in a position to tackle its medical problems: this amounts to a fifth of the countries in the world.[18]

That does not mean that the price of drugs is not a major problem. The World Health Organization estimates that there are over 300 essential drugs that should be available to all countries, but at least one-third of the world's population cannot afford them.[19] On the other hand, we cannot simply abolish patents. No new drugs will be developed if we do. The cost of developing a drug through all its stages until it is finally sold on the market can be up to $500 million. Even the most publicly spirited pharmaceutical company cannot afford to sell a patented drug at the price of a generic one. Several concessions have nevertheless been made by pharmaceutical companies for special cases. In the year 2000, five leading companies cut the price of anti-retroviral drugs used to treat patients with AIDS by 85 per cent and there have been further concessions for these drugs since then. Brazil has had the most effective policy of any developing country for limiting the effect of AIDS. By a mixture of local production of drugs, which are unpatented and cheap, and tough negotiations with the drug companies to reduce the price of the other retrovirals needed, it has been able to supply free treatment to over one-fifth of its 600,000 HIV/AIDS patients. The reduction in hospital admissions has more than compensated for the vast costs of the AIDS programme.

In several other fields there have been effective programmes of help initiated by pharmaceutical companies themselves. Merck has had a long-standing arrangement to supply its drug used to treat river blindness free of cost, helping towards the eradication of this disease; Pfizer makes a donation to Morocco of its drug for the treatment of trachoma worth $18 million a year. This programme too has been so successful that locally, trachoma is likely to be eradicated. NGOs such as Médecins sans Frontières argue that occasional acts of charity are no substitute for a generally fairer system. The profits of the pharmaceutical companies, they say, are high enough to enable them to take more account of need where this is greatest and the concessions are too limited. Even at the lower price, they argue, most countries would still be unable to afford anti-retroviral drugs.

On the other hand, the drug companies have a case for limiting concessions. Lower prices reduce the incentives that are badly needed, particularly for the treatment of AIDS. The AIDS virus shows a high rate of mutation, especially under the influence of drug therapy, and at a time when a stream of new drugs is required, the rate at which they are being developed is declining. There is a danger that pressures to reduce the price of necessary drugs may prove counter-productive by reducing their supply.

There are some tentative glimmerings of hope. It is possible to redress the bias that now exists against the development of drugs needed in the Third World. International organizations and governments can provide incentives to private companies to make such developments more profitable. The WHO, spurred on by its former director-general, Gro Harlem Brundtland, is taking a new lead in global health issues and the World Bank is increasingly turning its attention to health. Even individuals can make a difference, as Bill Gates has demonstrated with a $1 billion donation, through the Bill and Melissa Gates Foundation, for research and treatment of tuberculosis, AIDS, and malaria. A variety of bodies have been set up to develop new medicines for developing countries and to encourage smaller biotechnology companies to do basic research that may later be taken up by larger ones. Another encouraging development has been the success of the Rockefeller Foundation in persuading companies to give away patent rights that prevent the more rapid development of drugs and other new products specifically designed for the Third World. The next generation of GM products are likely to be more suitable for Third World needs than the first generation, which consisted predominantly of herbicide-tolerant crops.

Rich countries and big companies will also find that they have a commercial interest in improving the health of the world's poorest people. Disease is one of the principal causes of poverty, just as poverty is one of the main contributors to disease. As more countries lift themselves out of poverty, they will spend more money on health. More of their citizens will begin to suffer from the diseases of the rich, and, as they live longer, diseases of the old. Self-interest will dictate a change of priorities.

The role of multinational companies

Globalization in the eyes of its critics is synonymous with the rising power of international companies. These spread their investments and factories throughout the world, often through subcontracting and often at the expense of investment in their own domestic markets. They have become a force in almost all parts of the world, except large swathes of Africa. Company brands have become global brands. In the food and drinks industry, for instance, globalization has produced both greater variety and greater uniformity. On the one hand, supermarkets and restaurants in industrial countries offer a great variety of choice of exotic food and wines; on the other hand, American fast-food chains such as McDonald's and Burger King appear in identical form offering identical food in almost every major city throughout the world; people drink Coca-Cola or Pepsi fizzy drinks even in the poorest rural areas. Hollywood programmes dominate television screens, even in remote African and Asian villages. Everywhere multinational companies, overwhelmingly American, leave a large footprint.

Free trade is in some ways an abstract principle and the WTO is a relatively distant and nebulous organization. The most passionate protests of the anti-globalization movement are therefore directed at a more concrete and visible enemy, multinational companies, whose presence is ubiquitous and whose premises can be physically attacked in the world's major cities by the violent and anarchist wing of the movement. Multinational companies stand accused of using their power to subvert governments, forcing them to lower taxes and weaken environmental regulations, of exploiting cheap labour and even child labour, and of extending their domination of the world by the exploitation of brands and logos.

How powerful are multinational companies?

The charge that the top companies in the world are now more powerful than most national governments is based on a fairly simple error. Yet it is part of the litany of the anti-globalization movement, accepted without question by its leading figures, that 51 of the top 100 economies of the world today are multinational companies. To quote *No logo* by Naomi Klein,

By now we've all heard the statistics: how corporations like Shell and Walmart bask in budgets bigger than the gross domestic product of most nations; how, of the top hundred economies, fifty-one are multinationals and only forty-nine are countries. We have heard (or read about) how a handful of powerful CEOs are writing the new rules for the world economy.[20]

Precisely the same figures, that 51 of the top economies in the world are multinational companies, are quoted in virtually the same words in another bible of the movement, *The silent takeover* by Noreena Hertz, who argues that international companies are so powerful that national sovereignty has effectively been overthrown.[21]

When Klein writes that 'we' have all heard these statistics, she represents the views of an introverted coterie who rely on each other as independent sources to confirm each other's prejudices. The figures quoted are confused and are based on a fundamental economic misconception. As many commentators have pointed out, notably Martin Wolf and Jagdish Bhagwati,[22] GDP (gross domestic product) is a measure of value added, which cannot be compared with a company's sales. The value added of a company is the difference between the value of its sales and the value of its purchases. The budgets that Naomi Klein cites refer to company sales. To compare sales of companies with the GDP of countries either shows ignorance or is a deliberate attempt to distort and evoke prejudice. Even the most primitive traders do not confuse sales with profits value added or would quickly go bust if they did. In fact, the value added of the 50 biggest companies amounts to no more than 4.5 per cent of the value added of the 50 biggest economies. The claim is simply untrue.

Next, if the top companies were as powerful as their critics claim, they should be able to consolidate their power. But the league of big companies is constantly changing, a fact incompatible with omnipotence. None of the top ten companies in the world today, measured by market value, were in the top ten a decade ago. At that time Vodafone and Nokia, to cite two European examples, were unknown to the world at large. Even the largest international companies are subject to control through national regulation, but perhaps the most effective check that stops them exercising monopoly power is competition. Indeed, big companies are most powerful inside closed markets. Open borders weaken corporate power.

Multinational companies and national sovereignty

It is claimed that multinational companies use their power to compel governments to lower taxes. Not so. If the proposition were true, taxes should fall as globalization gathers pace. In fact the reverse has happened. In countries who are members of the Organization for Economic Co-operation and Development, the countries in which international companies have invested most heavily, the average ratio of tax revenue to GDP rose by over 5 per cent between 1980 and 1999, a very substantial rise. In the European Union, the ratio rose from 33.5 per cent to 42.3 per cent between 1970 and 1996. In the last three decades of the last century, Britain attracted more foreign direct investment than any other European country. If Klein and Hertz were right, the burden of corporate tax in Britain should have declined, certainly as a proportion of total taxation. Precisely the opposite happened. Between 1970 and 1996, while total tax receipts in Britain fell from 37 per cent to 35.6 per cent of GDP, corporation tax receipts rose from 3.1 per cent to 3.6 per cent, a figure appreciably higher than the EU average of 2.6 per cent.[23] Sweden and Denmark are countries that are among the most open to international trade and foreign investment, yet they have the highest tax rates in the world. In the end, tax rates and the levels of public spending are determined not by multinationals but by voter preferences. National sovereignty has not been overthrown.

Again, it is claimed that multinational companies seek out countries with the cheapest labour and stop governments passing regulations to prevent exploitation of labour. Not so. Most investment is not made in countries with cheap labour. As already mentioned, most American investment flows into Europe; companies invest in France, despite the enactment of a 35-hour week, and in Britain, despite the introduction of a minimum wage. As for investment in the Third World, of course there have been bad cases. Klein cites a number of examples where Nike, Adidas, Walmart, and other American companies have exploited cheap labour in the Philippines and even cheaper labour in China. In an appendix, she lists a rogues' gallery of companies and their sweatshops in China, making a strong *prima facie* case that there is widespread abuse.[24] Her book performs a service in drawing attention to malpractices and adding to public pressure to end them. But as a general argument against globalization, her case fails. Overall, foreign companies in poor countries pay higher wages than local employers; in Vietnam, for example, Nike's subcontractors (Nike are prominent villains in Klein's rogues' gallery) pay their employees double the average wage and provide much better working conditions.[25] In Indonesia, the average wage in a foreign-owned plant is 50 per cent higher than in domestic plants.[26] The International Institute of Economics in Washington, a highly respected independent think-tank, has also found that people in poor countries who work for foreign affiliates of American companies earn on average double the domestic manufacturing wage.[27] In poor countries, the lowest wages are paid in the local service sector, in small industries and farming, not in the factories, mines, or plantations of multinational companies. The reason foreign companies pay higher wages is that they need to attract labour of the highest quality. It is not therefore surprising that the Pew poll quoted earlier showed strong support in developing countries for the presence of multinational corporations.

It is worth adding that the use of child labour—and Klein cites several examples—is a more common practice in the world than most people realize. The International Labour Office estimates that in the year 2000, 186 million children aged 5–14 were in employment and 50 million aged 15–17 years.[28] In some cases, according to one of its officials, 'you have the absurd situation where the parents are unemployed and the children are working'.[29] However, to ban exports from companies that employ child labour, as some advocate,would simply increase poverty, which is the main cause for child labour in the first place. Ending the evil of child labour depends on the United Nations' commitment in the second of its Millennium Development Goals to universal primary education.[30] Globalization, by increasing general prosperity and reducing poverty, will help, not hinder, the achievement of this aim.

It is also claimed that multinational companies use their power to lower environmental standards. Not so. Most international investment has not been made in countries with least regulation. Furthermore, as globalization has spread, so have environmental regulations. If association proved cause and effect, this would indicate that globalization *increases* environmental protection. In fact, nation states favour stronger environmental regulation and national sovereignty prevails; big multinational corporations cannot prevent it, even if they wish to. Again, foreign companies in the developing world generally observe higher environmental standards than local companies.

Of brands and logos

The main theme of Naomi Klein's book, as its title indicates, is that multinational companies have used logos and brand names to extend and abuse their power. Like Klein, I favour the decentralization of power and, in principle, I support measures that reduce trends towards cultural conformity. I therefore sympathize with her aversion to forces that encourage people to develop the same tastes, wear the same clothes, and consume the same food and drink all over the world. (This is, however, what people want to do. No one is forced to drink Coca Cola against their will.) But she tries to have it both ways. Brand names have advantages too. Identification with brands has made the companies that own them highly conspicuous and therefore vulnerable to adverse publicity. Campaigns against sweatshops and child labour organized on American campuses would not have been successful if the brand names of the companies in question had not been universally recognized. Brands therefore provide some sort of safeguard against unethical practices. We cannot check up on the quality of every product we buy. Instead, we rely on brands, because they are a guarantee of quality: if a universally marketed product is proved faulty, it is a devastating blow to the reputation, and profits, of the company whose brand it carries, as Perrier discovered in 1990 when some of their mineral water was found to be contaminated with benzene.

Another point about brands, and indeed about the power of companies, that opponents of capitalism invariably underestimate: their influence is always subject to the effects of competition. Global brands tend to be more expensive than local products and command the high premium end of the market. As a result, local producers spring up to offer local products of only slightly inferior quality at a vastly lower price and usually better suited to local tastes and cultural preferences. Furthermore, the richer a country becomes, the more likely it is that brand choices will multiply. In the developed world, the number of kinds of soft drinks, of cars, mobile phones, cheeses, packaged foods, and own-label products on supermarket shelves is increasing all the time. It is not surprising that in 2000 some concluded that the sale of global brands had probably peaked.[31]

Capitalism and science

At the beginning of this chapter, I referred to the widely held belief that capitalism corrupts science and that research results produced by scientists who work for multinational companies should be disregarded, because the pursuit of profit cannot be reconciled with the public interest and such results are bound to be self-serving. As I have mentioned, there are many drawbacks about the growing dependence on corporate finance for scientific research and many examples of malpractices. Some scientists have been corrupted. Those who continue to contend on behalf of their corporate masters in tobacco companies that there is no link between smoking and cancer are an obvious example. On the other hand, most scientists who work for companies, including agri-businesses, are as likely to believe that their work is valuable and useful to mankind as those who work in universities or research institutes. Perhaps the biggest disadvantage of the decline of public sector research is that private funding directs research primarily into products that provide the highest return, not those most needed for the relief of poverty and hunger. The public sector funded the Green Revolution, and research into GM products most likely to benefit the poor and hungry is still largely funded by public money or by independent organizations such as the Rockefeller Foundation, and of course increasingly by public investment in research in countries like China, India, and Brazil.

However, these reservations do not invalidate the results of corporate research, because the validity of research findings is independent of the motive of the researcher. As argued in Chapter 8, the postmodernist view that what matters are the motives of scientists, since science is not value-free, is false. It is the quality of research that matters. If the researcher invents results to benefit a company, his or her reputation will be destroyed and the company's will suffer. Privately funded research can be good or bad; so can publicly funded research. Bad research will be discredited, however pure the motives of the scientist. Good research will prove its value, however selfish the motives of its corporate sponsor. No one can deny that many life-saving new drugs and other invaluable discoveries have come from the research laboratories of big corporations. Furthermore, the new generation of GM products, including new plants developed by agribusinesses, are likely to be particularly suited to Third World needs, and partnerships between the companies and philanthropic bodies like the Rockefeller Foundation should enable the new technology to be increasingly applied for the wider benefit of mankind.

As a general proposition, therefore, the view that capitalism corrupts science is not valid. Indeed a much greater threat to the independence of science comes from those who wish to assert political control and to make science subservient to some kind of ideology.

The values of capitalism

At the heart of the new anti-capitalist mood lies suspicion of the profit motive. You do not have to be an American neo-conservative to regard indiscriminate attacks on the profit motive as a naïve and dangerous over-simplification, if not distortion, of the role of business in society. Companies have virtues as well as vices and it is ludicrous to tar all businessmen with the same brush. Of course some businessmen will take the view that first and foremost they must maximize the profits of their company. For others, at least part of their motivation is to produce something useful to society. Indeed companies are no less accountable—some would argue they are more accountable—than the NGOs who demonize them. Both have special interests: companies to increase their profits, NGOs to increase their membership (which in the case of the Green lobbies depends on achieving maximum publicity). NGOs have to keep their subscribers happy, companies have to please their customers. If companies sell goods that are defective, their profits plunge. If biotechnology companies were to sell products that damaged human health, their profits would suffer grievously and they might even go bust. Distillers, the company responsible for sales of thalidomide, never recovered from the birth defects it caused. Indeed, it is the principal virtue of capitalism as a system that businesses only succeed if they satisfy customers. That was the reason why it proved more successful than socialism, which was based on a general and rather remote concept, the benefit of society, as the motivation for work.

Capitalism lacks grandeur. Marx had a vision for society that inspired those who wanted to change the world. Adam Smith offered no grand vision, but appealed to self-interest: 'It is not from the benevolence of the butcher, the brewer or the baker that we expect our dinner, but from their regard to their own interest.' There is nothing heroic or inspiring about self-interest. The beneficial qualities of capitalism are mundane: thrift, self-reliance, cautious investment, politely serving your customers, obeying the law, and paying your debts. Smith's invisible hand has been described as innumerable accountants balancing the books.[32] These are very pragmatic virtues, which fit in well with a Lockean but not with a Marxist approach to society.

However pedestrian the capitalist vision may be, it works and it is the most effective instrument yet devised for the production of wealth. It has one other great merit: an effective market economy is decentralized, which has both a political and an economic advantage. The political advantage is that decentralization militates against the concentration of power and dictatorship. The economic one is that it promotes experiment and innovation. The very nature of business under capitalism makes business people receptive to new ideas since innovation can be vital to success, unlike a career in the civil service, academic administration, or politics. It is no accident that in recent decades most Nobel Prizes in science were awarded to residents of the United States and that the Nobel laureates mainly work in independent institutions, whose trustees have a predominantly business background.

But perhaps the most important lesson which the anti-capitalist protesters forget is that whatever the excesses and abuses perpetrated by some capitalists, the only real alternative to the profit motive is state ownership, which generally means the death of enterprise and the asphyxiation of science and innovation. Green critics of capitalism also forget that centralized planning and universal state ownership not only stifled enterprise but caused the worst pollution in the industrial world.

The anti-globalization movement

What does the anti-globalization movement stand for? Can it offer, as some of its champions maintain, an alternative which is neither capitalist nor socialist and which avoids the mistakes of the past? What can this strange alliance, of Greens linking arms with trade union members representing smoke-stack industries, vegetarians allying with meat farmers, Trotskyites mixing with church groups, eco-fundamentalists joining with anarchists, possibly have in common, except what they are against?

Since the terrorist attack on the Twin Towers in New York on September 11th 2001, a World Social Forum has been set up that meets annually in Porto Alegre in Brazil, with the aim of transforming the anti-globalization movement from mere protest into something more positive. The task of founding a new political force is not a simple one. Marxism was based on hard thinking and much writing by Marx and Engels. The United States of America, one of the few countries founded on a constitution based on declared principles, officially came into being after long deliberations among a small number of learned and enlightened individuals. The European Union has recently sought to give coherence to its complex institutions and express its basic values in a new constitution. This was drawn up after many months of argument in a convention of some 100 parliamentary representatives from different member states. When new political parties are started, the first steps are taken by small groups of like-minded people who then submit the new party's proposed principles to a larger body for ratification. At Alegre, by contrast, as its organizers boast, over 100,000 people came together from all over the world to form a new democratic movement. At a meeting of the more modest offshoot of the World Forum, the European Social Forum, some 40,000 gathered, made up, according to a report in *The Guardian*,

of intellectuals, students, ecological and social activists, people representing the poorest and most marginalized, radical economists, concerned individuals, humanitarians, artists, culturalists, churches, scientists, and land workers from a bewildering array of non-government groups and grassroots social movements.[33]

In the circumstances it is not surprising that what has emerged so far is vague. At the first Alegre conference the emphasis was on the creation of a movement that was 'new', new faces, new ideas, new methods, and a determination to avoid the failures of left-wing regimes of the past. The second conference was dominated by big gatherings to hear speeches by big personalities, some of whom preached something uncomfortably close either to 'new' Marxism, or old-fashioned left-wing politics. President Hugo Chávez of Venezuela, for example, not at first sight a reassuring advocate of the democratic decentralization of power, declared that 'the left in South America is being reborn' and cited the continuation of Fidel Castro's rule in Cuba as evidence. Castro is clearly much admired by those who went to Alegre, and even if he owes much of his popularity to his ostracism by America, identification of the new movement with the Castros and Chávezes hardly suggests that its positive programme will be either democratic or new. Indeed, Naomi Klein declared her disappointment that the second World Social Forum was usurped by 'big men and swooning crowds', instead of building its own version of participatory democracy.[34]

Unfortunately, her own alternative prescription shows little awareness of the real world of politics or economics. As disclosed in a series of essays and articles, she envisages a movement based on neighbourhood councils, participatory budgets, stronger city governments, land reform and co-operative farming, referendums, constituents' assemblies, and empowered local councils, 'a vision of politicized communities networked internationally to resist further assaults from the IMF, the World Bank and World Trade Organization'.[35] Budget constraints, intellectual property rights, and multinational companies are conspiracies against the public that should be dispensed with. It is an eloquently articulated reaction against widely felt injustices, based on a conviction that everything is getting worse, that everywhere democracy is being trampled underfoot by monolithic capitalism, and that oppression and inequality in the world can be cured if only we abandon the 'neo-liberalism' of the market, put people before profits, and restore power to the people.

To point out that both analysis and prescription are simplistic and flawed is to underline the obvious. Many forms of decentralized politics mentioned by Naomi Klein have been widely advocated by liberal democrats in many countries, but they are neither inconsistent with, nor an alternative to, globalization. Democracies have to strike a difficult balance between local powers and central decisions (local centres of power may reach separate conflicting decisions that are nationally incompatible and can only be resolved centrally) and also between the will of the majority and the rights of minorities. It is naïve to denounce budget constraints and patent rights, to fail to realize that governments that ignore budget constraints can eventually go bust and that technological innovation requires some system of patent rights. However strong the case for the reform of TRIPS, immensely costly investment in research and development of new drugs will never be made if competitors can sell copies at a price that merely reflects the cost of production.

Nor is 'granting power to the people' a simple remedy to prevent democracy being trampled underfoot as globalization spreads. In fact, while there are still numerous unpleasant dictatorships around the world, their number is declining, not increasing. According to the latest Human Development Report, the number of regimes considered democratic jumped from 44 in 1985 to 82 in 2000, while the number considered authoritarian declined from 67 to 26.[36] This is a remarkable improvement, even if it must be conceded that democratic institutions in the new democracies are not yet very firmly based. In any case, democracy consists of more than the slogan 'power to the people', or Klein's plea (echoing Rousseau) that we must not oppose the repeated and stated will of the majority.[37] Stalin, Hitler, and Mao all had overwhelming popular support at some time in their careers.

If Naomi Klein's view of the movement's principles or programme lacks substance, at least it provides a wealth of detail compared with the declaration of aims of the European offshoot of Alegre—the European Social Forum—which proclaimed its new vision for Europe at a meeting in Florence in November 2002. Europe, it declared, should be an area with open borders; all those who live in Europe should have the right to work and to have a home, there should be no GM foods, no pollution, no racism, the media should be in the hands of the many not the few, and it should promote fair trade with poor countries. The virtues of motherhood were overlooked.

The anti-globalization movement is clearly not a movement based on reason. It represents an emotional reaction against power, authority, and technology, of 'us' against 'them'. Seattle 1999 was 'the precise and thrilling moment', wrote Klein, 'when the rabble of the real world crashed the experts-only club where our collective fate is determined'.[38] There will never be a positive, alternative programme, as she acknowledges in one of her essays, partly because of the way the movement operates. It is the response of the internet generation to the international economic system: these web activists 'have no top-down hierarchy to explain the master plan, no universally recognized leaders giving easy soundbites—and no one knows what is going to happen next'.[39] Its essence was perhaps captured by one of the slogans at Genoa: 'Capitalism should be replaced by something nicer'.[40]

This simple slogan demands a postscript. Just as it is wrong to tar all businessmen with the same brush, so all forms of capitalism cannot be treated as the same. The anti-globalization movement assumes that capitalism is uniform. It broadly equates it with 'the American Business Model', in which companies are left to create wealth in an environment of low taxation and a minimal role for the state, whose economic role is mainly confined to protecting property rights, and which allows greed to be the principal motivator.[41] This business model has undoubtedly been successful in creating wealth in the last decade, particularly for company directors, but its claim to be the most successful model in the world is open to question.

Capitalism was a European, not an American, invention and there are many different models on show. Even in the United States many companies behave in ways very different from those expected from the American model. Not only is 'social capitalism' in Europe very different from its American relative, but inside Europe itself the corporate system of almost every country is distinct. There is the German 'Rhineland' model (now undergoing significant alterations), the Dutch 'Polder' model (just as successful in the last decade in promoting growth and employment as the American model), a more *dirigiste* French model, separate Scandinavian models, and so on. Outside Europe, in Japan and Korea for example, there are also very different models. Nearly all the European models stress workers' rights, recognize a measure of industrial democracy, and seek to operate by consensus. The role of the shareholder, while gradually acquiring greater substance, does not have the pre-eminence it is accorded in America, and rewards for directors, while still handsome by almost any standard, do not begin to approach the stellar heights of those of their American counterparts. Takeovers, which often treat companies and their employees as if they were commodities to be bought and sold like chattels, are much rarer outside the Anglo-Saxon world. Indeed, champions of European-type social capitalism argue that, at its best, it reconciles the advantages of the market with the interests of a fair society.

European social capitalism is viewed by many Americans as sclerotic and inflexible, lacking innovation and enterprise, and ineffective at delivering economic growth. In time, they argue, everyone will have to copy the American model. It is not within the scope or purpose of this book to diagnose the strengths and weaknesses of different capitalist systems, but on past evidence it seems a reasonable forecast that no single model will prevail. If the test of economic success is the record, not of the last ten, but of the last thirty years, European social capitalism wins the prize. European economies have grown faster than the American economy, not vice versa. Even today, despite the recent surge in growth of American productivity, production per hour is higher in several European countries, including France, the Netherlands, Belgium, and former West Germany, while it is only fractionally lower in Austria and Denmark, all of them countries which prosper despite their high rates of taxation, thus disproving the argument that high taxes necessarily stifle enterprise and destroy wealth.[42] Annual production per worker is lower because, adopting a civilized approach, European workers take longer holidays and work shorter hours. Even Germany, which has not yet fully recovered from the enormous burden imposed by reunification and which faces the need for major structural reforms to overcome a recent spell of low economic growth, is hardly a country facing a major crisis. The German social model offers stability, security, fairness, a very high standard of public services and a civilized lifestyle, in fact the kind of social good, as Europhiles would argue, that economic growth should aim to achieve. However, past successes do not prove that European models of capitalism will prosper in the next thirty years if they fail to adapt to new circumstances and there is little doubt, as persistent high unemployment in several EU countries demonstrates, that this is a time when many traditional business practices in a number of EU countries will have to change.

The movement against globalization views capitalism as monolithic, because, like eco-fundamentalists and the back-to-nature movement, it seeks simple solutions to complex problems and prefers to talk about the general, not the particular. Its motives, justice for the poor and more equal treatment between nations, are admirable; its intellectual base is weak. It is another example of the triumph of emotion over reason. If it succeeds in limiting free trade, abolishing the WTO, and replacing contemporary capitalism, it will not make the world a better place.

10

Reason and Democracy

Throw reason to the dogs. It stinks of corruption.

> Slogan on the wall of the Ministry for the Prevention
> of Vice and the Promotion of Virtue in Kabul, 1998

So Two cheers for Democracy: one because it admits variety
and two because it permits criticism.

> E. M. Forster

A recurrent theme of this book is that the rejection of reason and the evidence-based approach, whether it is manifested in the advocacy of excessive caution, in the return to superstition in medicine, in the mystical approach of organic farming or the irrational, semi-religious opposition to GM crops, undermines a civilized society and ultimately democracy. But how close is the link between reason and democracy? I argued in Chapter 1 that it was no coincidence that the birth of modern science at the time of the Enlightenment coincided with the beginnings of modern democracy. The advocates of the new politics of pragmatism, tolerance and respect for human rights, notably John Locke, explicitly acknowledged their debt to science for overthrowing witchcraft, superstition and dogma and for teaching us to base knowledge on evidence rather than authority. The development of science and the new freedom of thought, together with the economic progress which science made possible, were at that time all interrelated and interdependent.

The other side of the coin is that the rejection of reason has historically been closely linked with autocracy. A very different philosophical approach from Locke's, based explicitly on the rejection of reason, was that of the eighteenth-century Swiss philosopher Jean-Jacques Rousseau. Rousseau insisted that our judgments should instead be based on feelings and natural instinct. He too advocated democracy, but a collectivist form of democracy, whose essence was the forcible imposition of the general will from which the individual had no right to deviate. Emphasis on the absolute sovereignty of the majority led to a very different political outcome. In the event Rousseau, the enemy of reason, proved to be the inspiration for the reign of terror and the more vicious ideologues of the French revolution and could also be regarded as one of the progenitors of the dictatorship of the proletariat. By contrast, Locke, the devotee of science, who championed the protection of individual rights and personal freedom, can be regarded as the father of modern liberal democracy. A glance around the world today reveals that science flourishes in countries where liberal democracy is enthroned and the rights of the individual are protected. Conversely, in countries where tyranny rules and individual rights are suppressed, science has made little headway.

Do democratic politics have a rational basis?

However, this coincidence between the success of science and the establishment of democracy does not in itself prove that democracy and reason are causally linked. In theory, there can be many other explanations for the coexistence of democracy and the success of science. Indeed, many will argue that modern democracy and its citizens are more noted for irrationality than reason and that Jonathan Swift was right to observe that 'the bulk of mankind is as well qualified for flying as for thinking'.[1] Obvious examples that support Swift's view are the popularity of astrology and homeopathy and of the national lottery, the fact that in the world's largest democracy, the United States, polls show that 49 per cent of the public believe in possession by the devil (another 16 per cent are not sure), 47 per cent in the account of creation given in Genesis, and about a third that people are periodically abducted by aliens.[2] Most people's daily activities are as likely to be governed by their horoscope as by the precepts of the scientific method.

At first sight, the electoral process is no more likely to inspire confidence in the importance of reason to democracy. The modern politician's stock in trade is the soundbite, which oversimplifies issues and empties them of content. My own memories of canvassing as a candidate in elections are of an exercise of mind-boggling superficiality. You shake people's hands or talk to them briefly on the doorstep and seek in one or two simple sentences to explain why they should vote for you rather than for your rivals. Much political debate consists of one party abusing another in exchanges of mindless partisanship. In the United States, money has become so important that the system of government sometimes resembles a plutocracy rather than democracy. Michael Blumberg became mayor of New York in 2002 after spending more money per voter on his election than any other candidate in history. Much of this money is spent on advertising, particularly short bursts of negative television advertisements that show an emphasis on personality rather than issues. If money and personality matter as much as argument, how can the process be described as rational? Dispassionate concern with reason is indeed the last quality the public in almost every country ascribes to politicians. In opinion surveys that rank people by the degree of trust they command, politicians come near the bottom—roughly equal with estate agents, debt collectors, and car salesmen, and only journalists come out worse.

One reason politicians behave the way they do is the attitude of the electorate. Because of public opposition, necessary or desirable reforms are frequently postponed or abandoned. In their memoirs many leading politicians have confessed that there were times when they knew what had to be done but could not do it because public opinion would not let them. As a result, governments have often won elections that they deserved to lose and lost when they deserved to win. If the electorate behaved rationally, more members of the public would vote to decide how they will be governed. In the United States less than half the population

normally vote in Presidential elections and in Britain the proportion of the electorate who vote in general elections is also dimin-
ishing.[3] As disillusionment with politics grows and membership of
political parties declines, so the influence of professional activists,
fund-raisers, and party officials increases and money plays an
increasingly important role. This trend towards plutocracy in turn
further increases cynicism about politics, because inevitably suspicion is aroused (and frequently justified) that those who pay the
piper call the tune.

Furthermore, on a closer look beneath the surface, a persuasive
case can be made that politics in democracies are about the clash
of interests rather than of ideas and that arguments between political parties about issues cloak the reality: elections are essentially a
competition between rival parties for the opportunity to promote
the interests of the groups they represent. Politics are therefore
about power. Sometimes the interests of particular groups prevail
even when they conflict with the interests of the majority. A good
example is the massive subsidies paid to farmers in the European
Union and the United States because they are powerful pressure
groups. These subsidies damage the national economies. What is
far more serious is that they also harm poor farmers in the rest of
the world and aggravate poverty in developing countries.

No wonder that Winston Churchill famously described democracy as the 'worst system of government except for all the rest'—
scarcely a ringing endorsement of democracy as an institution
founded on reason.

Why democracy is based on reason

Nevertheless, the role of reason in the democratic process can be
understated. To start with, politicians get a worse press than they
deserve. The public sees them through the distorting vision of
the media, yet most serious political commentators who know
politicians well do not hold them in low esteem. As one might
expect, individual politicians are no less (or more) responsible

than the run of people anywhere. In my own experience of politicians in Britain, I have found some outstanding for their intelligence, integrity, and courage, others to be pompous, self-important, and dishonest. Most are perfectly reasonable people. On the basis of a limited experience of politicians in the United States, I would accept the conclusion of a commentator on American affairs that 'politics in America is practised by a governing class whose members within the normal limits of human behaviour, mean more or less what they say and more or less keep the promises they make'.[4] In most mature democracies politicians do not go into politics to make money, or are fools if they do; a surprising number want to make the world a better place and all of them are ready to argue their case.

Most importantly, liberal democracies depend on tolerance and the need to compromise, and politicians do in fact compromise, listen to the other side, and are willing to modify their own position in the light of public discussion and public reaction. However reluctantly, political parties accept the verdict of the electorate when it goes against them. Successful democratic politicians are gradualists, not revolutionaries. Extremism and dogmatism may rear their ugly heads from time to time, but they do not win in the end.

As for the present widespread cynicism about politics in Britain, no one should underestimate the influence of the media. It makes good copy to suggest that politicians are all incompetent rogues. One famous editor of the *Sunday Times* and later *The Times* advised his journalists: 'Always ask yourself when interviewing a politician: why is that bastard lying to me?' As John Lloyd, himself a journalist, wrote: 'The modern journalistic imperative [is] Do Harm. If some metaphorical blood is not drawn, [an] interview is a failure'.[5] A good example of the readiness of the media to attribute dishonesty to politicians was the allegation made by a BBC reporter (given worldwide publicity) that Tony Blair and his advisers had deliberately misled the people about the information they had received about Iraq's weapons of mass destruction. An independent inquiry reported in 2004 that it found the allegations untrue. I was not surprised. It is rare for British Governments, whatever the party in power, to set out deliberately to mislead. However, it is significant that the report was immediately dismissed by most of the press as whitewash, on the grounds that no report which absolved politicians from the charge of lying could possibly be objective. When the media cross the line that separates scepticism, the essence of effective journalism, from derision, cynicism and ridicule, they no longer bolster democratic politics but bring them into contempt. The other side of the coin of cynicism is gullibility. If you trust nobody, and anyone's word is as good as anyone else's, you are as likely to believe charlatans as someone who is telling the truth.

However, even when full account is taken of misrepresentation through the media, pandering to prejudice, emphasis on personality and all the less attractive features of election campaigns, issues raised in elections are generally, in a very rough and ready way, judged by voters on evidence. Are schools better since the existing Government came into power? Are jobs safer, are prices lower than they would have been if an alternative government had been in charge? Who is more likely to improve the environment or reduce crime? Who has done more, or is likely to do more, to benefit one's family and friends? The parties and their leaders seek to win by argument and by citing evidence, even if the argument is often oversimplified and the evidence selective. Those who grossly distort tend to be exposed in time and, as John Stuart Mill pointed out, 'wrong opinions and practices gradually yield to facts and argument'.[6] In most countries there is also a growing demand for transparency and accountability to ensure more open discussion, in which the better case is more likely to prevail in the end. The electorate is a jury that pronounces a verdict on the record of the government in power and the claims of would-be alternatives. Like legal juries, it sometimes errs, but tries to base its verdict as best it can on the evidence presented.

Furthermore, even though most people subscribe to a number of irrational beliefs and are generally ignorant of science or the actual process of the scientific method, they are quite sensible

about many of the decisions that affect them directly. Most go to
qualified doctors when they are seriously ill. They expect the
police to act on facts not fictions. In fact a police investigation (if
properly conducted) is not unlike the process of scientific discovery. Detectives try to discover the facts, form a theory about
guilt, and test it against the evidence. They then have to convince
a jury (as a kind of peer review) that their conclusions are right.
Good garage mechanics also observe the scientific method. As
Robert Pirsig observed in his novel *Zen and the art of motorcycle
maintenance*, when the cause of a vehicle breakdown is uncertain, a
good mechanic will gather facts, formulate a theory, and carry out
tests to see if it stands up, quite unconsciously acting as any good
scientist would.[7] It recalls Locke's observation: 'God has not been
so sparing to men to make them two legged creatures, and left it to
Aristotle to make them rational'.

Democracy and the evidence-based approach

It is therefore not inappropriate to describe reason as one of the
cornerstones of democracy and irrationality as its enemy. And
even if the role of reason in politics is limited, democracy is the
only means through which the process of deciding by evidence
can find political expression, unlike dictatorships where fiat and
authority rule and choice is banned. Indeed a parallel can be
drawn between the ways in which democracies develop and the
way in which scientific knowledge grows, though there can, of
course, be no exact parallel. As Marx failed to realize, there are no
natural laws governing history or societies. Societies are made up
of people and people are unpredictable. There are also different
forms of democracy, which behave in very different ways. Nevertheless, relevant comparisons can be made, and democracy and the
evidence-based approach have important features in common.

Scientific knowledge has been defined as tentative knowledge:
scientific hypotheses start by being tentative. Some are then so
strongly confirmed by evidence that they become established facts.

For example, everyone accepts as true beyond doubt the laws of aerodynamics that enable aeroplanes to fly and sailing boats to go to windward and the laws of physics that stop bridges falling down. There is a consensus among scientists that Darwin's theory of natural selection is no longer a theory (whatever the creationists may say) but a true description of the way species evolved. But the scientific method itself involves critical examination and testing of every new hypothesis and many hypotheses will be replaced in time. Compare the way democracy evolves: three of its most essential elements are freedom to criticize, tolerance of different viewpoints, and a willingness to compromise. Dogmatic certainties have no place. Criticism, when it is justified, leads in time to the adaptation and improvement of democratic institutions which decay if they ossify. Indeed a key to the success of liberal democracies in the world is their capacity to allow change.

Karl Popper, who wrote one of the seminal books about the essential nature of democracy and its contrast with dictatorship,[8] also convincingly described the link between democracy and the scientific method.[9] The reason why countries with the highest living standards in the world are all liberal democracies, he argued, is not because democracy is a luxury which their wealth enables them to afford. The causal connection is the other way round: democracy is essential to the creation of higher living standards.[10] Societies progress by the free assertion of differing proposals, followed by criticism, followed by the genuine possibility of change in the light of criticism. Popper argued that all policies, indeed all executive and administrative decisions of government, involve empirical predictions about what will happen, which often turn out to be wrong. In fact it is usual for them to have to be modified in practice. Thus a policy is a hypothesis that has to be tested against reality and corrected in the light of experience. Mistakes are more likely to be avoided if there is critical examination and discussion beforehand. Furthermore, it is only by critical examination of the practical results that mistakes will be identified soon after they occur. Authoritarian institutions, by contrast, which forbid critical examination both before and after policies are decided, press on with mistakes long after they have begun to produce unintended and harmful consequences. The whole approach of an authoritarian society is anti-rational. A rational and scientific approach requires societies to be open and pluralistic.

Approaching the issue from a very different perspective, the distinguished economist Partha Dasgupta, who examined the quality of life in fifty-one of the poorest countries in the world, found that those citizens who enjoyed greater political and civil liberties during the decade of the 1970s also experienced greater improvement in life expectancy and in real income per head.[11]

The pluralism of democracies has been a vital element in their economic success. Indeed economic systems are embedded in their social and political context. In the light of experience, democracies adopted a market system in which economic decisions are decentralized, thereby allowing experiment and innovation and enabling people to learn from each other's mistakes and successes. As a result, innovation has flourished. Mistakes have not had the devastating repercussions of wrong policies universally applied, such as, for example, Khruschev's agricultural reforms in the Soviet Union in the 1960s.[12] The centralized authoritarian model of Communist states stifled innovation and prevented errors from being recognized at an early stage or limited in their consequences. That is why their economies could not compete with the Western democracies and finally broke up.

Autocracy and the rejection of the scientific approach

The Soviet Union

Another feature of both democracy and the evidence-based approach, which also links the two together, is that the enemies of the first are the enemies of the second. Ideologues and fundamentalists who do not allow evidence to sway them from their fixed beliefs and dogmas, who never compromise and who refuse to consider or try to understand any argument advanced by political opponents or unbelievers, not only reject the democratic process but also reject the scientific method. Science needs to breathe the air of free discussion and free criticism and is stifled by autocracy. In countries where ideologues have held or hold power, neither democracy nor science have prospered or prosper today. In countries where ideologues are an influential minority, their influence acts as a brake on progress and threatens the climate of tolerance on which a healthy democracy depends.

The Soviet Union was the classic case of a planned society based on ideology. It claimed that Marxism was 'scientific', because Marx felt he had discovered the laws of history and of capitalist production, which enabled him to predict human behaviour and the future development of society. Marxists designed a blueprint for society, which, they believed, could resolve the conflicts that plagued capitalism. In fact, Marxism cannot be described as scientific because there was no room under Marxism for tentative hypotheses that might be critically examined and discarded if evidence did not confirm them. Marxism was the true faith and criticism of the faith was treason punishable by death to the infidel. Apart from its belief in the fallacy that there are laws of history that correspond to laws of nature, Marxism was based on a denial of the basic elements of the scientific approach.

This was clearly demonstrated by the history of the life sciences in Soviet Russia under communism. It is a woeful chronicle, in which a few charlatans, playing cleverly on the rigid ideology that passed for thought within the ruling caste, succeeded in poisoning all science throughout the Soviet Union and its Eastern European imperium.[13] Lysenko ruled Soviet biology. He was a man with a hypnotic presence who came from a peasant background, an important advantage at the time, and took a degree in agriculture. He soon became lauded as a 'barefoot professor'. He invented a new technique for plant breeding called 'vernalization', a technique allowing winter crop seeds that had been chilled and soaked to be planted and grown in summer. This, he promised, would produce unheard-of abundance. He adopted Lamarckism, the discarded doctrine that acquired characteristics can be inherited, and when this doctrine was officially accepted by Stalin and the politburo, he ensured that every other thesis was ruthlessly suppressed. Most scientists who contested his views disappeared into the gulags or were forced to recant (like Galileo). Bourgeois, class-ridden genetics were banished. The doctrines of Mendelian inheritance were denounced as 'the ravings of a monk'. A new 'creative Darwinism' had to be adopted, based on dialectical materialism. This dictat was dutifully accepted by communist scientists in Britain and France as well as in the Soviet Union. A professor in the Faculty of Medicine in Paris, for instance, loyally declared that 'the principles of dialectical materialism [are] the most powerful tool in scientific thought'. Materialist Soviet science could allow no place for 'the idealist and mystical tendency founded by reactionary biologists, Weismann, Mendel, and Morgan'.[14]

Russian physics presented a more mixed picture. A number of leading physicists had acquired worldwide renown and the regime recognized the importance of physics to its purposes. Indeed they kidnapped a famous physicist, Pyotr Kapitsa, who worked in Rutherford's Cavendish laboratory in Cambridge, during a visit he paid to his homeland in 1934. But the leading physicists in due course came under savage ideological attack and it is likely that physics would have gone the way of biology had it not been saved by the atomic bomb and been allowed to flourish more or less unhindered. When ideologues at a later stage called a meeting to denounce them for their bourgeois concepts, Stalin cancelled the meeting and told Beria: 'We can always shoot them all later'. When the Soviet atom bomb was exploded in 1949, they were showered with honours. And when they faced renewed censure at the time of the alleged plot by Jewish doctors, many of them were saved by Stalin's death.[15]

Nazi Germany

The Nazis were even more dismissive of the scientific method.[16] When he came to power in 1933, Hitler declared:

We stand at the end of the age of reason ... A new era of the magical explanation of the world is rising, an explanation based on will rather than knowledge. There is no truth, either in the moral or the scientific sense ... Science is a social phenomenon, and like those, is limited by the usefulness or harm it causes. With the slogan of objective science the professariat only wanted to free itself from the necessary supervision of the state.[17]

Mussolini and the Italian fascists, although they did not subscribe to the full list of Nazi absurdities and racist doctrines, nevertheless had no time for a reasoned or scientific approach. Mussolini's message to Italians was simple: 'Believe, obey, fight'.[18] Truth was not highly regarded in Mussolini's Italy. Primo Levi, the Italian writer and chemist who was deported to Auschwitz and survived, declared science to be 'an antidote to the filth and fascism which polluted the sky':

the chemistry and physics on which we fed ... were the antidote to fascism ... because they were clear and distinct and verifiable at every step, and not a tissue of lies and emptiness, like the radio and the newspapers.[19]

Nazi philosophy was based on the doctrine of the blood, or 'blood and soil'(*Blut und Boden*). Just as Communists believed in a superior Communist biology, Nazis believed in a pure, transcendent Aryan physics. Some biologists in Nazi Germany went to great lengths to manufacture unsubstantiated hypotheses to justify the racism of regime. Thus, science was not proper science if it was practised by Jews. Heinrich Himmler, the powerful head of the Gestapo, was obsessed with astrology and the occult and founded a special German form of Creationism or *Welteislehre*, a theory about the creation of the world out of ice and fire that Hitler said he was also inclined to believe as correct. In the words of a chronicler of Nazi science: 'a torrent of bilge [flowed] from the academies of a country that so prided itself on the effulgence and rigour of its scholars'.[20] The Nazis seemed blind to the absurdities of their own doctrines. Julian Huxley pointed out that 'the ideal Nazi was to be as blond as Hitler, as tall as Goebbels [who was short and club-footed] and as slim as Goering [who was hugely corpulent]'.[21]

One theme that characterized biology in the Nazi era was holism in all its aspects, a theme still important to modern devotees of the 'Back-to-Nature' movement, especially those who advocate alternative medicine and organic farming. 'The approved *volkisch* medical doctrine was that all elements of the body were interconnected: there was thus no room for clinical specialists and their mechanical doctrines. Conventional academic medicine was replaced by herbs, homeopathy, sunshine and fresh air'.[22] The Back-to-Nature movement was much favoured by the Nazis. Indeed, it is interesting to note that Rudolf Steiner, the spiritual father of the organic movement, was a member of the party in the 1920s. The Nazis were also strong supporters of antivivisection, and the SS, as mentioned in Chapter 6, were taught to show the deepest respect for all animals. Altogether the values and philosophy of Nazism were a complete antithesis to everything the Enlightenment stood for: human rights, freedom of thought, egalitarianism, anti-slavery, anti-colonialism and above all, the evidence-based approach to knowledge.

Neither the Nazis nor the Soviet Union rejected the contribution technology could make to military purposes. In both countries rocket science was highly developed. However, neither state made any major contribution to scientific knowledge as a whole while ideology ruled. This was particularly significant in the case of Germany, which in pre-Nazi days secured more Nobel prizes per head than any other country, but whose science suffered a double blow: not only the official rejection of reason and the evidence-based approach by the Nazi ideologues, but the loss of the immense contribution made by Jewish scientists. Judged by the number of Nobel prizes it has won since the Second World War, German science has not yet fully recovered. Science in the United States, and to a lesser extent, in Britain, gained from Germany's loss.

The threat from religious fundamentalism

Today, the threat to democracy from ideologues is more likely to come from religious than political movements. In Europe, despite an occasional burst of support for the far right in France or Austria, extremist parties have limited influence and are not likely to come to power. In the United States, some of the more extreme pronouncements of neo-conservatives may evoke echoes from the McCarthy era of the 1950s, but it seems unlikely that freedom of speech will come under serious threat. On the other hand, religious extremism threatens peace and undermines democracy wherever it appears, whether in Asia, Europe, or America. In India, Sikh fundamentalists spread terror in Amritsar, Hindu fundamentalists destroyed the Babri Masjid mosque at Ayodhya, and Muslim terrorists have perpetrated a number of bomb outrages. Occasionally communal violence breaks out between extreme Hindu and Muslim groups with heavy loss of life. In Kashmir, Muslim and Hindu fanatics from time to time raise fears that there could be nuclear war between India and Pakistan. The Islamic fundamentalists of Al-Qaeda have declared a campaign of terrorism against the United States and its allies. Jewish fundamentalists occupy and seek to extend their occupation of Palestinian land, and are one of the biggest obstacles to peace in the Middle East. Finally, Christian fundamentalists support the fundamentalists in Israel and undermine the strong democratic traditions of the United States.

However, the relevance of fundamentalism to my theme is not its danger to peace, or to democracy alone, but its relevance to democracy and science. Fundamentalism was originally defined as strict adherence to orthodox tenets, held to be fundamental to the Christian faith (Oxford English Dictionary). In this sense, it requires belief in the literal truth of a sacred text, whether this is the Bible or the Koran, whether the fundamentalism is Christian, Jewish, or Islamic. In a broader sense, it means the passionate adherence to a set of convictions that are not amenable to reason:

they cannot be shaken by evidence and are unchangeable. An open exchange of views, on which all democracy and science relies, becomes impossible if minds are closed to argument or evidence. Islamic fundamentalism, in the form of Wahhabism or Salafism, presents a particular threat to democracy at the present time, partly because it is the most dynamic fundamentalism and most successful in winning converts in different parts of the world, and partly because in its most extreme manifestation, the Takfiris, it is associated with the complex and diverse network of terrorists of Al-Qaeda.[23]

In some quarters it is widely regarded as taboo to criticize any religion, even in its fundamentalist forms. We can freely abuse people for their political views, for their atavistic socialism, or their greedy capitalism, but religious beliefs are sacred and generally regarded as immune from criticism. As Douglas Adams, author of *The hitchhikers guide to the galaxy* said, 'If somebody thinks taxes should go up or down, you are free to have an argument about it. But if somebody says: 'I mustn't move a light switch on a Saturday', you say, 'I respect that'.[24] When Richard Dawkins suggests that belief in the physical translation of the Virgin Mary into Heaven (which was not mentioned in the Bible) shows a certain credulity and disregard for the laws of nature, his comments are widely regarded as in bad taste and a sign of intolerance. It seems quite acceptable for people to go on believing certain things at the cost of heaving all science overboard as irrelevant if it is part of their religious faith. No political commentator, as far as I am aware, has pointed out that one conclusion that seems to follow from recent acts of terrorism is that there would have been no attack on the Twin Towers and there would be fewer suicide bombers if it were not for a belief in an afterlife.

Hebrew fundamentalism

Consider first Hebrew fundamentalism. No rational doubts may challenge the absolute truth and divine authorship of the Torah and its completeness as the totality of revelation. In 2003 the chief rabbi in Britain, Jonathan Sacks, wrote a book in which he said:

> In the course of history, God has spoken to mankind in many languages: through Judaism to Jews, Christianity to Christians, Islam to Muslims . . . truth on earth is not, nor can it aspire to be, the whole truth . . . in heaven there is truths, on earth there are truths. No one creed has a monopoly on spiritual truth.[25]

This Lockean doctrine of tolerance was promptly condemned by a group of orthodox Jews as heresy, for daring to suggest that different religions have something to learn from each other. One rabbi cited Proverbs 10:7 'the name of the wicked will rot' and declared that the book should be recalled and all its copies destroyed. The book was duly withdrawn and rewritten.

Such claims to a monopoly of religious truth by the orthodox inevitably lead to deeds of intolerance. When no one may turn on a light switch on the Sabbath, the dictates of orthodox religion may seem absurd to non-initiates, but harmless. But when Jewish settlers in Israel cite Deuteronomy: 'For ye shall pass over Jordan to go in to possess the land which the Lord your God giveth you, and ye shall possess it, and dwell therein' and justify their seizure of Palestinian lands on the basis of holy scriptures written over two thousand years ago, fundamentalism becomes a cause of bloodshed and war. However, whatever the impossibility of reconciling such fundamentalist beliefs with modern science, and whatever one's views of the danger to peace caused by the occupation of Palestinian land by Jewish settlers, Jewish fundamentalism has not so far prevented the state of Israel itself from being a democracy. Nor has it interfered with the development of science in Israel, which has a high reputation. If the religious parties were ever to gain control, then the future of both science and democracy would be at risk.

Jewish fundamentalism has been greatly strengthened by the support it receives from Christian fundamentalists of the religious right in the United States. There are some 10–15 million evangelical Christians in America, often called Christian Zionists, who think it is contrary to God's will to put pressure on the Israeli Government. To them, concern for Israel is second in importance only to the fight against abortion. They essentially regard themselves as Bible-believing Christians, but according to one specialist monitoring the political influence of religion, they are perhaps best described as 'dispensationalists', because they believe that history should be divided into seven distinct eras or 'dispensations' and that we are now in the seventh era; this means that the end of the world is nigh.[26] They equate modern Israel with the Israel of the Old Testament, the descendants of Abraham to whom God gave Israel in perpetuity. As one evangelist declares on his website, 'We support Israel because all other nations were created by an act of man, but Israel was created by an act of God'. Therefore, there can be no Palestinian state in biblical Israel and of course Jewish settlements must remain. A body called Christian Friends of Israeli Communities, founded in 1995, runs a twinning programme between Israeli settlements and US churches.[27] Of course, if rights of occupation were to be based on history, perhaps 'native Americans' should be granted suzerainty over North America and aborigines should be given the right to reclaim Australia. Unfortunately for them, they have no ancient religious texts to justify retrospective re-occupation.

Christian fundamentalism

There is no doubt that in recent years religious groups in America have become an increasingly powerful force in the Republican party. Today, the religious right is far more powerful in American politics than it was one or two hundred years ago. Its influence is particularly strong in primary elections. They dictate policy on abortion and population control in the developing world. Their successes, combined with President Bush's declaration that he is an avowed 'born-again Christian', have led some people to argue that American foreign policy is now driven by religious lobbies. Fortunately, however, not all religious groups in America think alike about foreign policy. The Catholic church, for instance, takes a very different line on policy in the Middle East (though not on abortion). The fastest growing religion in America is Islam, and in time its political influence will also carry weight. Furthermore, the failure of the campaign by evangelical Christians to prevent China from being granted most-favoured-nation trading status shows their power is limited. It is also worth noting that several of the senior figures who decide the course of US foreign policy are more accurately described as 'Hobbesians rather than holy rollers'.[28]

Christian fundamentalism, because of its intolerance and dogma, co-exists uneasily with democracy and is incompatible with the evidence-based approach. However, the reason why both democracy and science flourish in countries where the Christian religion prevails is the fact that most Christians, even when they accept the authority of the bible as the literal word of God, also accept the separate jurisdictions of church and state. In the United States the constitution specifically guarantees it. Nor does religion dictate every aspect of life. Catholics, for example, in principle accept the authority of the Pope and his fiats about what they should or should not believe, but in practice the Pope's injunctions are frequently ignored. The Pope has ruled that contraception is immoral, yet Catholic countries in Europe such as Italy and Spain have some of the lowest birth rates in the world, about half that needed to maintain their present population. If present trends continue, in two-hundred-years time there will be only one Italian and one Spanish baby born for every sixty-four born today. Other groups accept that the Bible is the Word of God, but interpret the text in ways that avoid a direct conflict with science. The Evangelical Alliance in Britain, for example, has produced a balanced (and readable) report on genetic modification that effectively refutes the arguments of eco-fundamentalists.[29]

The separation of religion and politics does not apply to the issue of abortion, which often evokes extreme manifestations of intolerance and in the United States has even led to violence and murder. It therefore comes as a surprise to learn that the doctrine that abortion is a form of murder does not have biblical sanction. As the Bishop of Oxford, Richard Harries, pointed out in a debate about stem-cell research in the House of Lords in 2002, until amended by Papal Bull in 1869, official Catholic doctrine, based on Thomas Aquinas (who in turn followed Aristotle), stated that the soul did not enter the embryo on conception, but 40 days later in the case of the male, and 90 days later in the case of the female.[30] The passionate belief of the 'pro-life' lobby that abortion is murder is based on a religious interpretation of comparatively recent origin in the history of Christianity. However, even if this were more widely known, it seems unlikely to temper the zeal of the anti-abortion lobby.

Another disturbing feature of American society is the success of Christian fundamentalism in spreading the teaching of creationism. If more young Americans are taught creationism—or its euphemistic equivalent, 'intelligent design'—will American democracy continue to be based in the longer-term on the principles of reason and tolerance that inspired the founders of the US constitution, and will science continue to prosper? Creationism deliberately teaches children untruths. It is a fact that the earth goes round the sun, not vice versa, and that the earth is round, not flat; what creationists do not accept is that Darwin's original theory of natural selection has been so extensively confirmed by evidence that it can be regarded as a fact, in much the same category as the fact that the earth is round and goes round the sun. Creationists argue that the biblical record shows that the earth is a few thousand years old; fossil records and radioactive dating demonstrate that this is plainly wrong. Just as Calvin decreed that Copernicus' demonstrations must be ignored because they contradicted the Bible and the authority of Copernicus could not be placed above the Holy Spirit, so creationists tell us that our understanding of the process of evolution, perhaps the most important development in biological science in the last hundred and fifty years, must be dismissed because it conflicts with the declared word of God.

Belief in creationism in America cannot be dismissed lightly as a minor eccentricity since opinion polls show that about half the population declares its belief in the literal truth of the bible.[31] If creationists took over the national curriculum, it would be a disaster comparable to the take-over of Soviet biology by Lysenko. Children would be taught to turn their backs on the scientific method and the legacy of Locke and return instead to irrational doctrines that inspired the Crusades, the Inquisition, the burning of witches, and all the other forms of persecution that prevailed before the Enlightenment. Fortunately American democracy is one of the most firmly established democracies in the world and no doubt in time the tide will turn. Crankish views and religious eccentrics have always been part of the American scene. The American constitution also wisely provides for the separation of church and state and prohibits the teaching of religion in state schools. Advocates of the teaching of 'intelligent design' have to overcome a formidable constitutional hurdle.

Islamic fundamentalism

Islam presents a more formidable challenge, both to democracy and the scientific approach. In the present climate of anti-Muslim prejudice provoked by the destruction of the Twin Towers in New York in 2001, it may seem tactless and dangerous to good community relations to criticize Islam. Just as some Zionists call anyone anti-Semitic who criticizes the policies of the Government of Israel, so any suggestion that the lack of democracy and economic backwardness of many Muslim states is linked to the nature of Islam is regarded as a manifestation of intolerance towards all Muslims. This should not deter us from trying to look at the nature of Islam as objectively as possible.

In the last fifty years and more, Islam has been a failure as a political movement. Pan-Arab nationalism failed under Nasser in Egypt and later under the Ba'ath parties in Syria and Iraq. Many of the most tyrannical and poor regimes in the world today, and none of the more democratic and prosperous states, are Islamic countries. Most Islamic countries have also failed economically. Over the past quarter of a century, according to the Brookings Institution, GDP per person in most Islamic states has fallen or remained the same.

Yet Islam was once the fount of knowledge. In its golden age, between the eighth and thirteenth centuries, Arab thinkers invented trigonometry and algebra, laid the foundations of chemistry and modern medicine, and led the way in astronomy. Islamic countries were culturally diverse and other cultures were not seen as a threat but as an intellectual resource. The world of Islam was highly receptive to ideas from outside: Islamic medicine took ideas from the Greeks, Romans, Indians, and Persians. Islamic agriculture imported a variety of foreign crops, from rice and sugar cane to aubergines and spinach. Islamic libraries were filled with translations of foreign works. Thus in medieval times, in most of the arts and sciences, scholars in Western Europe were the pupils and the Islamic world provided the teachers. Historically, the followers of Islam were also more tolerant than Christians or other religions, allowing pluralism and religious freedom in the countries they controlled. They often sheltered Jews and Christians fleeing persecution in the West (for example, the Ottoman Empire welcomed Sephardic Jews after their mass expulsion from Spain in 1492). Moreover, in its early days, the Islamic world was more liberal than contemporary Christian societies in its attitude towards women.[32]

Why, then, did the birth of modern science after the Renaissance take place in Western Europe rather than in Islamic countries, which had for many centuries been more advanced, richer and more enlightened? Why has democracy failed to establish itself in Islamic countries today, and why have they failed to emulate the economic progress, not only of the West but of the vibrant economies of Asia?

Sometime in the fourteenth century, Islam changed and the Muslim world abandoned scientific inquiry. By the end of the century observatories that were unique in the world had disappeared, partly for religious reasons. When printing presses were invented in Western Europe, they were banned by Islam because the printed word might spread undesirable material that would undermine their faith. The great cultural changes of the West, the Renaissance, Reformation, and the Enlightenment, went apparently unnoticed in the lands of Islam and the new scientific literature of Europe was almost unknown to them. Until the late-eighteenth century only one medical book was translated into a Middle Eastern language, and that was a sixteenth-century treatise on syphilis.[33] A report prepared for the United Nations Development Programme by a group of academics pointed out that since its early days there has been little Arab translation from other languages. In the almost 1200 years since the death of the Caliph Mamoun in 833, the Arabs have translated as many books as Spain translates in a single year.[34]

Orthodox Islam

Vicenzo Oliveti, a European expert on Islamic studies, groups the House of Islam today into three major ideological divisions: Orthodox, Fundamentalist, and Modernist.[35] Orthodox Islam has a traditional canon, a collection of sacred texts which everyone agrees are authoritative and definitive, and which has fixed the principles of belief, practice, law theology, and doctrine throughout the ages.[36] The canon starts with the Koran itself and includes the great traditional Commentaries upon it, the eight traditional collections of the sayings of the prophet Mohammed, and in addition a number of other texts and sources which scholars have consulted and interpreted over the years to work out their practical applications and details. Orthodox Islam, unlike the extremists in the Fundamentalist division, does not support violence, does not seek to force its religion on others, does not prescribe the oppressive treatment of women or force them to wear black veils, and does not desecrate historic works of art or statues from other cultures. It is opposed to the violence preached by Al-Qaeda.

Yet Orthodox Islam cannot be absolved from blame for the backwardness and undemocratic nature of modern states in which Islam is the main religion. As Oliveto, who clearly has great admiration for Orthodox Islam, points out, the religion of Islam is far more pervasive in the daily lives of Muslims than Christianity is in the lives of Christians.[37] Many more attend the mosque on Fridays than Christians attend churches on Sundays. Most Muslims pray five times a day. Islam is more prescriptive of behaviour than Christianity, including prescriptions about what to eat and how to dress. One reason freedom of thought became established in the West was that the role of the church and of religion became less pervasive. In Christian states, church and state became separated, a recognition that you render unto God what is God's and unto Caesar what is Caesar's. This separation was a vital factor in the development of democracy, as was the separation of religion and the law: democracy requires that the lay people's representatives make laws. The Renaissance, the Reformation, and the Enlightenment changed the nature of society, broke the stranglehold of the mediaeval church, and liberated a spirit of inquiry that allowed democracy to develop. Under orthodox Islam no part of life is outside the scope of religious law and jurisdiction. Philosophy is 'the handmaiden of theology and science merely a collection of bits of knowledge and devices'.[38] There is no separate church and no clergy (except in Iran since the seizure of power by Ayatollah Khomeini). The state is the church and the church is the state and God is head of both, or, to put it another way, 'Islam, the ruler, and the people are like the tent, the pole, the ropes and the pegs. The tent is Islam, the pole is the ruler, the ropes and the pegs are the people. None can thrive without the others'.[39]

Oliveti contrasts Orthodox Islam with fundamentalist Islam: the former tolerant, preaching peace; the latter intolerant and oppressive, but also militant with a strong violent wing. However, it can also be argued that all devout Muslims are in a sense 'fundamentalists', in that they believe in the 'fundamentals' of their religion and consult the holy texts, seen as authentic and revealed, for guidance.[40] The belief that there is only one law that is of divine origin, the sharia, and that the Koran was the text revealed to Mohammed, delivered to him by Allah and thus perfect, unchanging and unchangeable, has deprived the Islamic world as a whole of the freedom to question and inquire and of the freedom from indoctrination, that are all essential to the development of science and of democracy. Because it constantly seeks to determine the correct reaction to contemporary events by reference to historic texts, it is a backward-looking religion. As Oliveti acknowledges, not one technological invention or scientific breakthrough in the whole twentieth century originates from Islamic countries,[41] and the blame for this failure cannot be attributed to the group he terms the Fundamentalists, since they have only recently acquired power and influence. The reason is that there is no room for scepticism in the Islamic religion itself. Apostasy is traditionally the most serious crime, punishable by death. Even today, few who were born Muslims question the faith itself. It is common for people to lapse from Catholicism and other forms of Christian faith in large numbers; not from Islam. Most Muslim countries are profoundly Muslim in a way most Christian countries are no longer Christian.

Modernist Islam

The third division of Islam today is Modernist Islam, born in Turkey and Egypt at the start of the last century from a concern that Islam was being left behind technologically by the West and needed to be updated. Today, modernist reformers aim to overhaul Islamic law and doctrine and they question the literal interpretation of the Koran as the immutable word of God. They advocate a secular state, in which religion is confined to the domain of the personal, one that can reconcile Islam and democracy. The modernists can point to the growing number of Muslim scientists in good Islamic universities in India and California, as well as to recent developments in Turkey which suggest that there is no incompatibility in principle between Islam and the modern democratic state. Yet at present modernizers are swimming against a strong tide, because the events of the last two decades, in Afghanistan, Iraq, and above all, Palestine, together with American policy since the destruction of the Twin Towers, have radicalized Islam and strengthened the appeal of the extremists. There is now a far greater sense of the *umma*, the Muslim community, than at any time since Western colonial powers broke up the remains of the Islamic empire eighty years ago.[42] The best hopes of Modernist Islam depend on the success of democracy in Turkey. If Turkey, a nation that has secularized Islam, can meet the democratic requirements for membership of the European Union and share in its prosperity, it will weaken the forces of tradition, reaction, and oppression elsewhere in Islamic countries. This is a powerful argument for supporting Turkey's application to join the EU. The alternative, the triumph of traditional Islam, let alone of Salafi fundamentalism, would be the triumph of unreason and autocracy.

Religious fundamentalism is not an important force in Europe, although it should be mentioned that the British Government recently licensed schools that teach creationism and, like some countries on the continent but unlike the United States, also provides state support for religious schools for historical reasons. Such schools treat children not as the children of Protestants, Catholics, or Muslims, but as Protestant, Catholic, and Muslim children. The difference matters, because it assumes that children should not make up their own minds about religious beliefs, but will automatically adopt those of their parents, or should be indoctrinated at school to ensure that they do. Moreover, the excuse for it, that it allows each community to preserve its own culture, in effect places religious and theological authorities in substantial control of some ethnic minority groups and narrows intellectual horizons. Although the quality of teaching in these schools is generally high, their influence is likely to be divisive and promote intolerance, as separate religious schools have done in Northern Ireland.

Eco-fundamentalism and democracy

Whereas movements of religious fundamentalism in Europe are weak, the forces of eco-fundamentalism are strong, and the stronger they are, the greater the threat they pose to democracy. Some people will feel that to accuse eco-fundamentalists of endangering democracy is to make a mountain out of a molehill. Does it matter if passionate Green activists exaggerate, have a cavalier approach to evidence, use tendentious language and propaganda, or see conspiracies around every corner? So do many other special interest groups, including partisan adherents of political parties. Even if views based on ignorance of or hostility to science lead to the spread of homeopathy and organic farming and stop farmers in Europe cultivating GM crops, why should this particular form of irrationality constitute a threat to our democratic system?

Most controversial issues are at least partly matters of opinion. In economic arguments, for example, it can sometimes be difficult to find two economists who agree. But scientific issues are different. New scientific theories are also disputed, but in time they are resolved because unlike economic theories or political credos they can be verified or falsified objectively. When there is a conflict of views between peer-reviewed papers published in a reputable scientific journal such as *Nature*, written by scientists recognized as experts in their field on the one hand, and a highly partisan document, not peer reviewed, published by campaigning bodies like Greenpeace or the Soil Association on the other hand, to reject the former in favour of the latter is to renounce reason as a basis for judgment. Yet how is the public to judge between conflicting scientific views, when it does not appreciate the importance of peer review, finds it difficult to tell a crank from a good scientist and is anyway suspicious of expertise? ('They told us BSE was safe'). Moreover, unorthodox opinions may sometimes be proved right, although we should remember that for every Galileo there are ten thousand Duesbergs (the doctor who claimed that there was no link between HIV and AIDS). The public needs help in deciding whom to trust, but the media, its main or only source of information about science, often misleads.

With the exception of the specialist scientific correspondents (whose reports are generally of high quality, but who are often ignored when a major story breaks) most journalists are almost as ignorant of science as the public. They also have a natural sympathy for mavericks. When the author of the study that led to headlines about 'Frankenstein food', Dr Arpad Pusztai, was forced to retire from his research institute because of his unprofessional behaviour, he was presented as a martyr and hero, who defied the establishment to warn the public. That his experiments were discredited by every reputable body that examined them went largely unreported by those who had first used his findings to promote the scare. Similar unbalanced reporting affected the views of the public about the safety of the mumps, measles, and rubella vaccine. Dr Andrew Wakefield, the doctor who claimed, in 1998, that the MMR vaccine causes autism, also acquired heroic status as an anti-establishment whistleblower. When scares are raised, no doubt the media cannot avoid reporting them, but they should at least issue a caveat if the alarm is not supported by a significant body of evidence or by most scientific opinion.

Many journalists feel an instinctive sympathy with Green pressure groups because they share their concern about the environment. Journalists are more likely to question the motives of corporations or those financed by them, than those of campaigning NGOs, although they too have their own agenda. Furthermore, the media and NGOs have a common interest in scare stories which increase circulation for newspapers and the membership of campaigning groups. Another important factor is the superb skill displayed by NGOs in media manipulation: no one manages public relations better than Greenpeace. Every story about a new development concerning transgenic crops or any article in favour of GM crops is followed by an instant comment from Greenpeace and its allies, often with an added gloss suggesting a sinister development. Until I started writing about controversial issues like organic farming and GM crops, I had not realized how strong and well organized the Green campaigners are. The media have largely adopted the language of the Green NGOs: for example, there are frequent references to the danger of 'contamination' from GM crops—a word which suggests corruption, pollution, and perhaps poisoning—when there is no evidence at present that cross-pollination from transgenic crops causes any environmental damage. This bias in favour of Green pressure groups is disturbing because it is clear that on scientific issues the public depends for its information on the media. In the case of the MMR scare, for example, an opinion poll showed that 20 per cent of the public thought there was a possible link between the MMR vaccine and autism, while 53 per cent thought the arguments for and against a link were evenly balanced.[43]

It is therefore clear that through the influence of the media, Green activists have succeeded in giving the public a skewed picture of several controversial scientific issues. Still, the question remains: does it all really matter? On a practical level it is clear that it can cause damage. In Ireland, where the anti-MMR campaign also ran strongly, there was a substantial rise in the incidence of measles and some children died. The future of agricultural biotechnology in Britain, in which we have great scientific strengths, is now bleak as a result of the campaign against GM crops. This campaign has also undermined public faith in an important branch of science and in the whole evidence-based approach. Together with media support for organic farming and alternative medicine, irrational fears are fostered about health issues that could easily break out into near hysteria if there was a serious epidemic—for example of influenza. The British public is not immune to local outbreaks of hysteria, as publicity about paedophiles has shown, even leading to an attack on a paediatrician. Newspaper scares can be dangerous: misinformation about asylum seekers could engender a mood that could destabilize a civilized society. Pre-war racial hysteria brought a Fascist government to power in Germany. Democracy depends on the supply of information that makes it possible for us to form independent judgements. If the public is swayed by demagogues, incited by jingoists, or inflamed by ethnic or religious chauvinism, democracy suffers. The UK is not about to see a surge in racism, but every step away from reason is a step on a perilous road.

The eco-warriors have encouraged a general cynicism about government and authority, have encouraged the public to suspect widespread corporate conspiracies against the public good and have added to the widespread suspicion that already exists of almost every kind of expertise. Criticism of authority is an essential and healthy part of democracy and unquestioning acceptance of expert opinion offends every independent spirit. But the assumption that all government and all industry is corrupt, when this is not true in most advanced industrial democracies, leads to the mindless nihilism of the anti-globalization protestors, who are eager to destroy but have no constructive alternative to offer.

Finally, fundamentalists who close their minds to ideas that conflict with their beliefs stifle the spirit of free inquiry. Those who exalt intuition, instinct, and feeling and promote them as the guiding principles of our society follow where Rousseau led, not what the Enlightenment taught us. They deprive us of our most powerful weapon, a rational vision of society. By championing the doctrine of unreason they seek to remove one of the cornerstones on which Western civilization is built.

I do not assume that my country is the best country in the world or that Western Europe or the United States or other older established democracies have a monopoly of wisdom or virtue, but I am not a cultural relativist. I do not believe we should undervalue the merits of Western democracy and civilization. Its merits do not depend on being Western, but on the nature of the society that happened for historical reasons to be founded in Western Europe, and which was itself based on the advances in science and learning made by the Islamic world in mediaeval times. Many of its values have now been adopted in other parts of the world. In the case of scientific discoveries, what matters is not where they were made or who made them, but what they are. Likewise, the merits of liberal democracy are independent of time and place and of the different countries that helped to shape it. Democracy crucially depends on the freedom to criticize, on rejection of dogma, and respect for evidence rather than authority. In the past it was also an expression of optimism about the world, a belief that progress is possible, that the condition of mankind can be improved. The scientific method, which advances our ability to improve our lot as it steadily expands our knowledge of the world, is also an expression of optimism. The scientific method and democracy are natural allies and unreason is their common enemy.

Epilogue

SINCE the dominant theme of this book has been exaltation of the evidence-based approach and excoriation of irrationality, it may invite the angry riposte of the philosopher George Moore in Stoppard's play *Jumpers*, when his wife suggests the church is a monument to irrationality:

The National Gallery is a monument to irrationality! Every concert hall is a monument to irrationality!—and so is a nicely kept garden, or a lover's favour, or a home for stray dogs. You stupid woman, if rationality were the criterion for things being allowed to exist, the world would be one gigantic field of soya beans!

(Had Stoppard written the play thirty years later, he might have made it a field of genetically modified soya beans.)

Of course, the George Moore character was right. Without poetry and music and love and laughter and even the thousand and one absurdities and trivialities of life that have nothing to do with reason, such as sport and even reality TV shows, life would be a desert. But to advocate the evidence-based approach is not to aspire to a world that is the slave of reason. The world of the mind is not the same as the world of reason; art and literature may have only an accidental connection with reason. Furthermore, discussions about social issues and politics, about justice and equality and the political good, may be highly rational but are rarely based on measurable results of reproducible experiments. The argument of this book is not that only arguments which are evidence-based are valid but that we should never ignore evidence where it is relevant. Even when it is relevant, I do not argue that evidence is all that matters.

For example, a wise philosopher may ask the question: 'What is it worth to gain the plaudits of the multitude if you lose the respect of your friends?' This is a question to which evidence might be relevant: there may be evidence from sociological research that tells us that the pursuit of fame and the plaudits of the multitude do not bring happiness. Juvenal wrote his classic seventh satire on the vanity of human wishes to show that fame was a false god. Dr Johnson asked (and answered) the question What did Charles XII of Sweden achieve by his victories and conquests in various parts of Europe? 'He left the name, at which the world grew pale, To point a moral, or adorn a tale'. Research may show that private citizens with many friends find life more fulfilling than celebrities with very few real friends, or that people are happier once they have given up public life for private life. Indeed, there is persuasive evidence that we are highly social animals who need friends and family to find true contentment. But even if some objective measurement of happiness could be devised and if this showed, unexpectedly, that fame was the most truly satisfying aim in life, it would not resolve the argument. There is also a value judgment to be made about the deeper quality of life, which cannot be based on a verifiable or falsifiable proposition. Not everyone's value judgment will be the same.

My main purpose, therefore, has not been to make exaggerated claims about the scope for applying the scientific method, but to wage war on those who ignore evidence, or the need for it, where an issue can only be decided by evidence. How can we tell if a herbal medicine is effective if people deny the need to test it and simply rely on the fact that it has been used for centuries? Many superstitions have survived for centuries. Whether organic farming is a good or bad system is for the most part a question capable of scientific proof: can people tasting the same fruit or vegetables grown in exactly the same conditions tell in blind tests which tastes better, those grown organically or by other methods of farming? Is organically grown food more nutritious or healthier for us? There are objective tests that can provide the answer. If organic food fails these tests, it is no excuse to plead, as the Soil Association does, that the tests are irrelevant because we must take the 'holistic' and spiritual dimension of organic farming into account. Some NGOs do themselves a disservice by their misuse of evidence about genetic modification, seizing on the flimsiest of experiments in their favour and sweeping all contradictory evidence, however weighty, under a rug. Criticism of GM crops by Greenpeace should not be taken seriously when its chief executive admits that *no* evidence can change its opposition. Sociologists who talk of the need to take into account 'unknown unknowns' in assessing risk should be laughed out of court. The nihilists who say there is no such thing as truth and the fundamentalists who believe that words written in sacred books a thousand and more years ago prevail over the discoveries of science are all enemies of reason—some of them very dangerous enemies indeed.

What matters most, however, is not who is anti-science or why, but what science has to offer, why being against it is to undermine the quality of the good society and why being indifferent to science is to be blind to one of the glories of mankind. It is the current mood of pessimism about science that I most deplore. There were good reasons why the birth of modern science inspired optimism and there are good reasons why it should still do so today.

It is the source of our knowledge of the world around us. How can it be argued that science is a dangerous assault on the integrity of nature, when it enables us to find out more about it? Those who know more about music enjoy it the more. Those who have studied art gain more intense pleasure than the rest of us from great works of art. The cricket lover who appreciates the subtleties of the bowler's wiles and the beauty of an effortless off-drive will find a day at Lords more rewarding than someone who only knows about baseball. To watch a flight of geese is a pleasure at the best of times, but how much more so for those who know about the great migrations of birds from continent to continent. The beauty of science is that it extends our knowledge, about ourselves, about the stars, about our whole environment, and that each scientist builds on the knowledge of his or her predecessors. Science breeds optimism because its province is knowledge, knowledge that improves and expands from generation to generation. The nature of science identifies it with the belief that progress is possible.

As a non-scientist, I envy scientists because I can only share

vicariously in the excitement of their discoveries. Some ignor-
amuses talk about science as if it means dull burrowing for facts,
when it does nothing of the kind. It is, as Karl Popper said, 'one of
the greatest spiritual adventures man has yet known'. A new
hypothesis is an act of creativity that is different from the creativity of the arts, because new creations of the scientific imagination
have to survive a detailed confrontation with experience and a
rigorous testing by sceptical critics, but it is an act of creativity
nevertheless. What is more, the discoveries of science make the
world a better place.

They do so firstly because the statements of science are tentative. The scientific approach is the enemy of absolute certainty. The
purveyors of certainties, the ideologues and the fundamentalists,
are the enemies of democracy. A tentative approach to knowledge
accepts criticism and breeds tolerance. Criticism and adaptability
are the characteristics of societies that are free and prosperous.
'Two Cheers for Democracy', wrote E. M. Forster, 'One because it
admits variety and two because it permits criticism.'

Secondly, because science and technology create wealth and
improve our quality of life. (Forster might have added this as a
third cheer.) Some environmentalists may despise the high living
standards of the western world on the grounds that higher consumption creates pollution and because many of the values of
consumerism are hollow and tawdry. But it is somewhat patronis-
ing to despise what nearly everyone in the world wants or to
dismiss the enormous benefits brought about by scientific medicine.
Greater wealth pays for better health care and better education
and enables people to live fuller lives (and makes it easier to control pollution.) More people may indulge in pointless luxury and
watch drab and worthless television programmes as society
becomes wealthier, but more people will also go to concerts
and art galleries and read books and live healthier lives. There
is no virtue in poverty, hunger, and disease, and science and
technology have done more to diminish them than any other
human discipline.

Finally, I come back to the link between science and progress. If by yielding to the view that science and technology create more problems than they solve, you give up hope of progress and desire for progress, you give up on civilization. Progress is not inevitable. It will not happen if the pessimists and the anti-science brigade prevail. From time to time, the human race lapses back into barbarism, as when the Hitlers, Stalins, Maos, and Pol Pots take over. But then it recovers and progress is made once again because there are people who believe in it and make it happen. Even if there are parts of the world in which tyranny still rules and misery prevails, there are now huge tracts of the world in which it has been shown that life can be better, free from oppression and from the worst stresses of poverty. Modern liberal democracy gives more people the chance of a good life than ever before. This would not be possible without the contribution of science.

Sources

PROLOGUE

1. Mayer Hillman (1992). *Cycling and public safety.* British Medical Association.
2. Jay Griffiths, quoted in *Resurgence*, Issue 174, January/February, 1996, p. 11.
3. Bjorn Lomborg (2001). *The skeptical environmentalist.* Cambridge.
4. R. M. Skirvin, F. Kohler, H. Steiner, D. Ayers, A. Laughnan, M. A. Norton and M. Warmund (2000). The use of genetically engineered bacteria to control frost on strawberries and potatoes. Whatever happened to all of that research? *Scientia Horticulturae*, **84**, 179.
5. Patrick Moore, Environmentalism for the 21st century. www.greenspirit.com
6. Mother Theresa was not above criticism. Apart from opposing birth control and contraception and spending most of the money she raised building convents rather than helping the poor, she refused pain-killers to those in her care because she believed that suffering pain was receiving the kisses of Jesus.
7. Jared Diamond (1998). *Guns germs and steel.* Vintage, p. 412.
8. T. Blair, *Science matters.* Speech to the Royal Society, 23 May, 2002.

CHAPTER ONE

1. Isaiah Berlin (1956). *Introduction to the age of the Enlightenment*, p. 29 New American Library.
2. Roy Porter (2000). *The Enlightenment, Britain and the creation of the modern world.* Penguin.
3. Quoted in Roy Porter, *op. cit.*, p. 6.
4. *Ibid.*, p. 8.
5. *Ibid.*, Introduction, p. xxii.
6. T. Holcroft, *Life of Thomas Holcroft* (1816 edn.; see Porter *op. cit.*, p. 508, note 27).

7. Porter, *op. cit.*, p. 70.

8. Quoted in Bryan Magee, *The story of philosophy*, Dorling Kindersley, p. 67.

9. John Henry (2003). *Knowledge is power: How magic, the Government and an apocalyptic vision inspired Francis Bacon to create modern science.* Icon Books.

10. See generally Simon Schama (1987). *Embarrassment of riches*, Harper-Collins, for a vivid account of Dutch society in the 16th and 17th centuries.

11. John Locke (1690). *Essay concerning human understanding, Epistle to the reader.*

12. *Ibid.*, p. xvii 4.

13. *Ibid.*, p. xix 1.

14. *Ibid.*, p. xvi 4.

15. *Ibid.*, p. xvii 5.

16. Isaiah Berlin (1991). *The crooked timber of humanity*, p. 237 Fontana.

17. For a variety of polling questions and answers about attitudes to science, see the House of Lords Report, Science and Society, 2000, Appendix 6, pp. 84–92. I have throughout this book frequently quoted from the reports of Select Committees of the House of Lords. They are one of the most important contributions of the upper house, as they hear evidence from leading experts in the issue under investigation and are an invaluable source of knowledge and are highly regarded in informed circles. Of equal value are similar reports from Select Committees of the House of Commons.

18. Anna Bramwell (1989). *Ecology in the 20th century.* Yale University Press.

19. *Ibid.*, p. 200ff.

20. George Steiner, *Times Literary Supplement*, 9 October, 1981. Heidegger was also one of the progenitors of the existentialist movement. See p. 197 below.

21. A. J. Lieberman and S. C. Kwon (1998). *Facts versus fears, American Council on Science and Health* (www.acsh.org). For a good summary of the effect of *The silent spring*, see Julian Morris (1999). *Fearing food*, Butterworth-Heinemann, Introduction, pp. xvi–xviii.

22. Julian Morris, *op cit*, p. xvii.

23. A. Wildavsky (1995). *But is it true? A citizen's guide to health and safety issues.* Harvard University Press; quoted in A. Trewavas (2004). A critical assessment of farming-and-food issues, to be published in *Crop Protection*, p. 5, Elsevier.

24. Rachel Carson. *The silent spring*, p. 13, Boston, Houghton Miflin.

25. John Maddox (1972). *The doomsday syndrome*, p. 110, Macmillan.

26. Anna Bramwell, *op. cit.*, p. 25.

27. Rachel Carson, *op. cit.*, p. 257.

28. M. V. Nadel (1971). *The politics of consumer protection*, p. 41, Bobbs Merrill.

29. S. Masterson-Allen and P. Brown (1990). Public reaction to toxic waste contamination. *International Journal of Health Services*, 20 (3), 487–97; quoted in Adam Burgess (2003) *Cellular phones, public fears and a culture of precaution*, p. 237, Cambridge University Press.

30. *Ibid.*, pp. 489–91.

31. R. Dworkin. *Playing God*. Prospect, May, 1999.

CHAPTER TWO

1. In the following summary of early developments in medicine, I have drawn extensively on Christopher Wanjek (2003). *Bad medicine*, Wiley.

2. Raymond Tallis (2004). *Hippocratic oaths*, p. 15, Atlantic Books.

3. Elaine Shapiro (2001). *The powerful placebo: From ancient priest to modern medicine*. Johns Hopkins University Press.

4. For a graphic description of the various treatments and their effects, see Wanjek, *op. cit.*, pp. 7–10.

5. UNDP Human Development Report, 2000.

6. A detailed examination of complementary and alternative medicine was conducted by a Select Committee on Science and Technology of the House of Lords, 6th Report, 2000. A balanced view of the merits and problems of alternative medicine can be found in the evidence from the Royal Society at pp. 43–51. (For the use of reports of Select Committees of the House of Lords, see Chapter 1, Note 17.).

7. *Ibid.*, Professor Meade, giving evidence on behalf of the Royal Society Q 179.

8. Professor Bateson, House of Lords, *op. cit.*, Q 175.

9. See Wanjek, *Bad medicine*, p. 169.

10. *Ibid.*, p. 175.

11. A large-scale clinical trial of the efficacy of acupuncture in treating migraine carried out in Germany showed that it was only effective as a placebo therapy.

12. Raymond Tallis, *op. cit.*, p. 129.

13. R. P. Feynman. *Surely you're joking, Mr Feynman*, pp. 338–9. Unwin Paperbacks 1185.

14. See Wanjek, *op. cit.*, p. 171.

15. P. R. Gross, N. Levitt, and M. W. Lewis (eds.) (1996). *The flight from science and reason*, p. 183. New York Academy of Sciences.

16. National Survey of Access to Complementary Health Care, Sheffield University, cited in Raymond Tallis, *op. cit.*, p. 128.

17. *British Medical Journal* (2001) 322, p. 181, cited in Tallis, *op. cit.*, p. 128.

18. *Ibid.*, p. 128, quoting the *Observer* supplement, 29 September, 2002.

19. Sheffield Survey, *op. cit.*

20. *Journal of the American Medical Association* (1998), 280, 784–7, quoted in Tallis, *op. cit.* p. 128.

21. House of Lords, *op. cit.*, Professor Meade Q 164.

22. Richard Dawkins (2003). *A devil's chaplain*, p. 180, Weidenfeld and Nicolson.

23. L. Dossey (1995). *Alternative Therapies 1, 2*: pp. 6–10, *The flight from science and reason, op. cit.*, p. 195.

24. Robert Park (2000). *Voodoo Science*, p. 53, Oxford.

25. House of Lords, *op. cit.*, written evidence, p. 210.

26. Daniel Moerman (2003). *Meaning, medicine and the placebo effect.* Cambridge.

27. 'A Trial of St John's Wort (*Hypericum perforatum*) for the Treatment of Major Depression', National Institutes of Health, National Center for Complementary and Alternative Medicine, 9 April, 2002.

28. K. Linde, G. Ramirez, C. D. Mulrow, A. Pauls, W. Weidenhammer and D. Melchart (1996). St John's wort for depression—an overview and meta-analysis of randomised clinical trials. *British Medical Journal*, 313, pp. 253–8.

29. G. di Carlo, F. Borrelli, E. Ernst, and A. A. Izzo (2001). St John's wort: Prozac from the plant kingdom. *Trends in Pharmacological Science, 22*, pp. 292–7.

30. *The Guardian*, p. 8. Health Supplement 4 February, 2003.

31. B. P. Barrett, R. L. Brown, K. Locken, R. Maberry, J. A. Bobula and D. D'Alessio (2002). Treatment of the common cold with unrefined Echinacea. *Annals of internal medicine*, 137. pp. 939–46.

32. NIH study, *op. cit.*

33. House of Lords Committee, *op. cit.*, Q199.

34. *Ibid.*

35. Roy Porter, *The greatest benefit to mankind: A medical history of humanity from antiquity to the present*, 1997, p. 712, Harper Collins.

36. Tallis, *op. cit.*, pp. 136–7.

37. Since this book was written a detailed history of the MMR controversy has been published: Michael Fitzpatrick, 2004, *MMR and Autism*, Routledge.

38. Tallis, *op.cit.*, p. 115.

39. B. Taylor, E. Miller, C. P. Farrington, M.-C. Petropoulos, I. Fauot-Mayaud, Jun Li and P. A. Waight (1999). Autism and measles, mumps, and rubella vaccine: no epidemiological evidence for a causal association. *The Lancet*, 353, pp. 2026–9.

40. The Guardian, 2 October, 2004.

41. Paul Offit, J. Quarles, M. A. Gerber, C. J. Hackett, E. K. Marcuse, T. R. Kollman, B. G. Gellin and S. Landry (2002). Addressing parents' concerns. *Pediatrics*, 109, pp. 124–9, January, 2002.

42. Simon Wessely, quoted in 'Myths of Immunity' www.spike-online.com

43. The effect of the 'Body parts Scandal' has been fully described in *The New Scientist*, pp. 14–16, 1 February, 2002.

44. *The Guardian*, 1 February, 2003.

45. *New Scientist, op. cit.*

46. *Ibid.*

47. Tallis, *op. cit.*, p. 189.

48. Robert Park, 31 January, 2003 (www.butterfliesandwheels.com)

49. Wanjek, *op. cit.*, p. 173.

CHAPTER THREE

1. Report of Policy Commission on Farming and Food, 2002.

2. See www.foodstandards.gov.uk/science/sciencetopics/organicfood/

3. ASA Adjudications, 12 July, 2000 www.asa.org.uk/adjudications

4. Quoted in Bryan Magee (2003). *The story of philosophy*, p. 156, Dorling Kindersley.

5. A. J. Trewavas, A critical assessment of organic farming-and-food assertions with particular respect to the UK, to be published, *Crop Protection* (2004), Elsevier. This paper is a comprehensive review of the issues concerning organic farming.

6. Anna Bramwell (1989). *Ecology in the twentieth century*, p. 200, Yale University Press.

7. House of Lords Select Committee, Organic Farming and the European Union, 1999, oral evidence Q 38. (The hearings of this committee provide a treasure trove of evidence about the arguments for and against organic farming.).

8. James Duncan, Letter, *Nature*, **425**, p. 15 (4 September, 2003).

9. Survey by *Health Which?*, quoted in House of Lords, *op. cit.*, p. 17.

10. A. J. Trewavas, *op. cit.* p. 4.

11. House of Lords, *op. cit.* Q462.

12. H. Hansen, Comparison of chemical composition and taste of bio-dynamically and conventionally grown vegetables, *Qualitas plantarum – plant foods for human nutrition*, **30**, pp. 203–11 and D. Basker (1992). Comparison of taste quality between organically and conventionally grown fruits and vegetables, *American Journal of Alternative Agriculture*, **7**, pp. 129–36.

13. Aventis Crop Science UK, January, 2001. See also Higginbotham *et al.* (2000). Environmental and ecological aspects of Integrated, organic and conventional farming systems, *Aspects of Applied Biolology*, **62**, pp. 15–20.

14. Joanna Blythman 'Toxic Shock', *Guardian Weekend*, 20 October, 2001.

15. John Krebs, *Nature*, **415**, 117 (2002).

16. Bruce Ames, *Fearing Food*, p. 25 (ed. Morris and Bate), Butterworth-Heinemann, 1999.

17. D. Coggon and H. Inskip, Is there an Epidemic of cancer?, *British Medical Journal*, **308**, pp. 705–8.

18. Trewavas, *op cit.*, pp. 5–6.

19. Hormesis has of course no connection whatever with homeopathy and its supposed law of infinitesimals. Firstly, there is no suggestion that 'like cures like'. Secondly, the low concentrations are measurable. They have not been diluted to some infinitesimal amount (of 1 to the power of 30 or more. See p. 44 above.).

20. See for example, Jaworowski, Radiation Folly in *Environment and health, myths and realities*, p. 68, ed. Okonski and Morris, 2004.

21. *Ibid.*, pp. 7–8 and E. J. Calabrese and L. A. Baldwin (2001). Hormesis: U-shaped dose responses and their centrality in toxicology. *Trends in pharmacological sciences*, **22**, pp. 285–91; (2002). Applications of hormesis in toxicology, risk assessment and chemotherapeutics. *Trends in pharmacological sciences*, **23**, pp. 331–7; (2003). Toxicology rethinks its central belief. *Nature*, **421**, pp. 691–2; J. Kaiser (2003). Sipping from a poisoned chalice. *Science*, **302**, pp. 376–9.

When toxicologists determine the safe dose of a chemical, they do so by characterizing its dose-response relationship, that is, how the change or degree of harm it induces varies with the different amounts present. They then determine the safe dose and how great a safety factor they should set. Generally there are doses at which the relationship is linear, but there are arguments about whether there is a threshold dose below which the chemical has no effect, or whether even one molecule induces a damaging effect, even if too small to be detectable, and the public has been led to believe that there is no safe level of exposure to many toxic agents, especially to carcinogens like radiation or dioxins. In fact it depends upon the mode of action of the substance in question. Biochemists and pharmacologists know that the dose response to many compounds, e.g. to sex hormones, opioids, antibiotics, anti-viral agents and non-steroidal anti-inflammatory drugs is U-shaped, in that small amounts have a stimulatory (or 'good') effect and large amounts are inhibitory (or bad). Ionizing radiation at low doses cures cancer, though higher doses induce cancer. Toxicologists now recognize that the same is true for many environmental chemicals regarded as poisons or as carcinogenic, in that very small amounts of arsenic, cadmium, mercuric chloride, aluminium or methanol for example, at doses below those which cause detectable damage are actually beneficial. However, the subject arouses controversy.

22. House of Lords, *op cit.*, Q 462.

23. Aventis Crop Science UK, January, 2001. See also S. Higginbotham, A. R. Leake, V. W. L. Jordan, and S. E. Ogilvy (2000). *Aspects of Applied Biology*, **62**, pp. 15–20 *et al.*

24. C. J. Drummond (2000). *Aspects of Applied Biology*, **62**, pp. 165–72.

25. Trewavas, *op. cit.*, p. 17.

26. Higginbotham *et al.*, *op. cit.*

27. Trewavas, *op. cit.*, p. 10.

28. Trewavas, Urban Myths of Organic Farming (2001). *Nature*, **410**, p. 409.

29. Trewavas, *op. cit.*, p. 9.

30. Dr. C. S. Prakash is Professor in Plant Molecular Genetics and Director of the Center for Plant Biotechnology Research at Tuskegee University. He was formerly on USDA's Agricultural Biotechnology Advisory Committee and is at present on the Advisory Committee for the Department of Biotechnology for the Government of India.

CHAPTER FOUR

1. Yet there are many examples of food on sale in the shops that has a GM content, or in the production of which genetic modification has been used. Thus, cheese sold as vegetarian cheese is made using the

enzyme chymosin, which is obtained from genetically modified micro-organisms instead of from calves' stomachs. Several food products contain soya from the genetically modified crop and many foods contain GM colourings, processing aids, vitamins and flavours. From April 2004 all except the processing aids have to be labelled if the amount of GM material exceeds 0.9 per cent of the total. Much livestock, hence meat, has been fed GM soya. In 2002–2003, a total of 36.5 million tonnes of soya (as beans or meal) was imported into Europe, of which about two-thirds was probably genetically modified. It is fed to livestock as a high-protein source that replaces the meat and bonemeal that was banned after the advent of BSE.

2. The survey was carried out by Cardiff University, the University of East Anglia and the Institute of Food Research. A copy can be obtained from A. Lopata (a.lopata@uea.ac.uk).

3. Alan McHughen (2000). *Pandora's picnic basket*. Oxford.

4. What the Monsanto European advertising campaign in 1998 actually said was: '. . . many of our needs have an ally in biotechnology and the promising advances it offers for our future. Healthier, more abundant food. Less expensive crops . . . With these advances we prosper; without them we cannot thrive. As we stand on the edge of a new millennium, we dream of a tomorrow without hunger. . . . Worrying about starving future generations won't feed them. Food biotechnology will' (quoted in 'Feeding or Fooling the World—Can GM crops really feed the hungry?', Report of the Genetic Engineering Alliance 2002, p. 13: see www.fiveyearfreeze.org). Monsanto's pitch overstates the case, but is not exactly the worst example of advertising hype that has ever been published.

5. Four separate reports by The Royal Society between 1999 and 2002, and most notably '*Transgenic plants and world agriculture*' a report by the Royal Society and the Brazilian, Chinese, Indian, and Mexican Academies of Sciences, the National Academy of Sciences, USA and the Third World Academy of Sciences, July, 2000.

6. Nuffield Council on Bioethics, 'The use of genetically modified crops in developing countries.' January, 2004. This was an update of an earlier report: 'Genetically modified crops: ethical and social issues', 1999.

7. ActionAid 'GM crops—Going against the Grain', May, 2003.

8. Nuffield Council on Bioethics, *op. cit.*, p. 37.

9. ActionAid, *op. cit.*, p. 17.

10. Nuffield Council on Bioethics, *op. cit.*, p. 53.

11. www.fiveyearfreeze.org

12. W. Stanley, B. Ewen, and A. Pusztai (1999). Effect of diets containing genetically modified potatoes expressing *Galanthus nivalis* lectin on rat small intestine (1999). *The Lancet*, **354**, 1353 its agricultural land and one-third of its forests, but these figures about soil erosion are challenged, with impressive evidential support, by B. Lomborg (2001). *The skeptical environmentalist*, pp. 104–6, Cambridge.

13. The Royal Society, GMOs and Pusztai, May, 1999.

14. In the above account, I have drawn extensively on an illuminating paper by A. J. Trewavas FRS and C. J. Leaver FRS (2001). *EMBO Reports*, **21**, p. 458. See also www.gmscience debate.org.uk/

15. T. Malthus (1798). *An essay on the principle of population.*

16. B. Heap FRS, Essay in B. J. Ford (ed.) (2003). *The scientists speak.* Rothay House.

17. Heap contends that in the past fifty years we have lost one-fifth the world's topsoil, one-fifth of its agricultural land and a third of its forests, but these figures have been challenged by B Lomborg, *The skeptical environmentalist*, pp. 104–6, Cambridge, 2001.

18. N. Myers (2000). Sustainable consumption, the meta-problem, in *Towards sustainable consumption: A european perspective*, pp. 43–8.

19. P. J. Gregory *et al.* (2000). Environmental consequences of alternative practices for intensifying crop production. *Agriculture, Eco-systems and Environment*, **88**, pp. 279–90.

20. Paul Ehrlich once famously wrote: 'The battle to feed humanity is over. In the course of the 1970s the world will experience starvation of tragic proportions—hundreds of millions of people will starve to death'. In Paul Ehrlich (1968). *The Population Bomb.* Ballantine Books, p. xi. Ehrlich's book sold 3 million copies.

21. News release from ISAAA (International Service for the Acquisition of Agro-biotech Applications), 13 January, 2004.

22. GM Crops, An International Perspective on the Economic and Environmental Benefits, p. 1–2, Gordon Conway, to be published in Van Emden and Gray, *GMOs an international perspective*, Elsevier.

23. B. Heap, *op cit.*

24. G. Conway, *op. cit.*, p. 5.

25. *Ibid.*, p. 5.

26. Center for Global Food Issues, 30 January, 2003 (www.cgfi.org).

27. G. Toenniessen, J. C. O'Toole and J. DeVries (2003). Advances in bio-technology and its adoption in developing countries, *Current opinion in plant biology*, **6**, pp. 192–3.

28. Dow Jones Commodities Service, New Delhi, 27 October, 2003.

29. *New York Herald Tribune*, 8 October 2003.

30. Toenniessen, J. C. O'Toole and J. DeVries, *op. cit.*, p. 194.

31. Nuffield Council on Bioethics, *op. cit.*, p. 16.

32. G. Conway, *op. cit.*, p. 16.

33. Colin Tudge (2003). *So shall we reap*, p. 268, Penguin.

34. G. Conway, *op. cit.*, p. 19.

35. Jonathan Rauch, *Atlantic Monthly*, October, 2003. pp. 105–6.

36. Nuffield Council on Bioethics, *op cit.*, pp. 35, 36.

37. The case for genetically modified crops with a poverty focus. H. J. Atkinson, J. Green, S. Cowgill, A. Levesley (2001). *Trends in biotechnology*, 18, pp. 91–6.

38. C. Tudge, *op. cit.*, p. 271.

39. Nuffield Council on Bioethics, *op. cit.*, p. 15.

40. *Ibid.*, p. 42.

41. Florence Wambugu (1999). Why Africa needs agricultural biotech. *Nature*, 400, pp. 15–16.

42. Broom's Barn. See Press release, 4 February, 2003. Broom's Barn, Higham, Bury St Edmunds, Suffolk, IP28 6NP, UK.

43. G. Conway, *op. cit.*, p. 8.

44. Conservation Technology Information Center, October 2002 (www.ctic.purdue.edu/CTIC/Biotech.html).

45. G. P. Robertson, E. A. Paul, and R. R. Harwood (2000). Greenhouse Gases in Intensive Agriculture: Contributions of Individual Gases to the Radiative Forcing of the Atmosphere. *Science*, 289, pp. 1922–5.

46. G. Conway, *op. cit.*, p. 945. G. Conway and G. Toenniessen (1999). Feeding the world in the twenty-first century. *Nature*, 402, C55–C58.

47. BBC News Online, 5 January, 2004.

48. H. S. Mason, H. Warzecha, T. Mor, and C. J. Arntzen (2002). Edible plant vaccines: applications for prophylactic and therapeutic molecular medicine. *Trends in Molecular Medicine.*, 8 (7), pp. 324–9.

49. G. Giddings, G. Allison, D. Brooks, and A. Carter (2000). Transgenic plants as factories for biopharmaceuticals, *Nature Biotechnology*, 18, p. 1151; J. Ma, P. Drake, and P. Christou (2003). The Production of Pharmaceutical Proteins in Plants. *Nature Reviews Genetics*, 4, p. 794; R. Petersen and C. Arntzen (2004). On Risk and Plant-based biopharmaceuticals, *Trends in biotechnology*, 22, p. 64.

CHAPTER FIVE

1. Colin Tudge (2003). *So shall we reap*, p. 254, Penguin.

2. Gordon Conway, GM Crops—an International Perspective on the Economic and Environmental Benefits, to be published in Van Emden and Gray, *GMOs an international perspective*, Elsevier, *op. cit.*, p. 5.

3. *Nature Biotechnology* (2003) **21**, pp. 1003–9.

4. ActionAid 'GM crops—Going against the Grain', May, 2003. p. 23.

5. T. Netherwood, S. M. Martín-Orúe, A. G. O'Donnell, S. Gockling, J. Graham, J. C. Mathers and H. J. Gilbert (February 2004). Assessing the survival of transgenic plant DNA in the human gastrointestinal tract. *Nature Biotechnology*, **22** (2), pp. 204–9. See also J. J. Flint *et al.* 'The survival of ingested DNA in the gut and the potential for genetic transformation of resident bacteria', www.botanischergarten.ch/debate/ Flintetal.pdf; and D. A. Jonas, I. Elmadfa, K.-H. Engel, K. J. Heller, G. Kozianowski, A. König, D. Müller, J. F. Narbonne, W. Wackernagel and J. Kleiner (2001). Safety considerations of DNA in food. *Annals of Nutritional Metabolism*, **45**, pp. 235–54.

6. John Vidal, *The Guardian*, July 17th, 2002.

7. T. Netherwood, *op. cit.*

8. Alan McHughen (2000). *Pandora's picnic basket*, pp. 119–21, Oxford.

9. Royal Society, February, 2002, *op. cit.*

10. New Scientist, Vol. 175, p. 7, 14 September, 2002.

11. Tudge, *op. cit.*, pp. 269, 346.

12. A. J. Trewavas and C. J. Leaver (2001). Opposition to GM crops: science or politics. *European molecular biology organization (EMBO)*, **21**, p. 459.

13. *Ibid.*, p. 458.

14. M. J. Crawley, S. L. Brown, R. S. Hails, D. D. Kohn and M. Rees (2001). *Nature*, **409**, 262.

15. OTA New Developments in Biotechnology (1988), p. 17.

16. Trewavas and Leaver, *op. cit.*, p. 459.

17. *Nature* (2002) **414**, pp. 541–3.

18. *Nature* (2002) **416**, p. 600.

19. H. Daniell, M. S. Khan and L. Allison (2002). Milestones in Chloroplast Engineering. *Trends in Plant Science*, 7, p. 84.

20. Editorial ('Terminator come back') (2002). *Nature Biotechnology*, **20**, p. 203.

21. *Reuters*, 28 January, 1999.

22. R. H. Phipps and J. M. Park (2002). Reduced pesticides. *Journal of Animal Food Science*, **11**, p. 1.

23. Conway, *op. cit.*, pp. 8–9.

24. Australian Associated Press, 27 October, 2002.

25. Conway, *op. cit.*, p. 9.

26. A. M. Shelton and M. K. Sears (2001). *The Plant Journal*, **27**, pp. 483–8.

27. *Ibid.*, p. 483.

28. J. E. Losey, L. S. Rayor and M. E. Carter (1999). Transgenic pollen harms monarch larvae. *Nature*, **399**, p. 214.

29. Shelton and Sears, *op. cit.*, p. 485.

30. Trewavas and Leaver, *op. cit.*

31. Shelton and Sears, *op. cit.*, p. 483.

32. Tudge, *op. cit.*, p. 258.

33. Trewavas and Leaver, *op. cit.*

34. Trewavas (1999). *Nature*, **402**, p. 231.

35. J. L. Fox (2003). *Nature Biotechnology*, **21**, pp. 958–9.

36. *Nature Biotechnology* (2003). **21**, pp. 1003–9.

37. http://www.rothamsted.bbsrc.ac.uk/broom/gm_work.html

38. Royal Society (2003). *Philosophical transactions*, **358**, pp. 1779–99.

39. ActionAid 'GM crops—Going against the Grain', May, 2003, p. 6.

40. *Ibid.*, p. 18.

41. M. Ceasar 'Transgenic Nation', *Bolivian Times*, 21 September, 2000, pp. 1, 3.

42. Ingo Potrykus (March 2001). Golden rice and beyond. *Plant Physiology*, **125**, pp. 1160–354.

CHAPTER SIX

1. Anna Bramwell (1989). *The history of ecology in the 20th century*, p. 204, Yale University Press.

2. Onora O'Neill (2002). *Autonomy and trust in bio-ethics*. pp. 78–82, Cambridge.

3. Anna Bramwell, *op. cit.*, p. 93.

4. Carolyn Merchant (1980). *The death of nature: Women, ecology and the scientific revolution*, p. xvi, HarperCollins.

5. Michael Crichton, Speech to the Commonwealth Club, San Francisco, 15 September, 2003.

6. Rory Spowers (2003). *Rising Tides*. Canongate.

7. *Ibid.*, pp. 350, 351.

8. *Ibid.*, p. 208.

9. *Ibid.*, p. 226.

10. D. Quinn (1999). *Beyond Civilisation.* Harmony Books; quoted in Spowers, *op. cit.*, p. 331.

11. *Ibid.*, e.g. pp. 262, 275, and 311.

12. *Resurgence*, Issue 199, March/April, 2000, p.17, quoted in Spowers, *op. cit.*, p. 311.

13. *Ibid.*, p. 287.

14. Quoted from the environmentalist David Suzuki, in Spowers, *op. cit.*, p. 125.

15. *Ibid.*, p. 84.

16. *Ibid.*, p. 220.

17. *Ibid.*, p. 248. See also pp. 233, 237, 240, 305, 343 and *passim*.

18. www.greenspirit.com

19. New Scientist, Vol. 164, p. 74, 25 December, 1999.

20. House of Lords, EC Regulation of Genetic Modification in Agriculture 1999 Evidence. p. 43, Q107.

21. John Emsley. *The consumers' good chemical guide*, p. 173, Oxford University Press.

22. *Ibid.*, p. 203.

23. Will Hively, Discover, Vol. 23, no. 12, December 2002.

24. B. Lomborg (2001). *The skeptical environmentalist.* Cambridge.

25. W. Beckerman and J. Sekal, *Justice, posterity and the environment*, p. 71, OUP, 2001.

26. Green Futures, 2004, Number 47, pp. 19ff.

27. World Commission on Environment and Development, 1987, p. 43.

28. W. Beckerman, *A poverty of reason*, The Independent Institute, 2002, p. 13.

29. Quoted in W. Beckerman (2001). Economists and Sustainable Development. *World Economics*, 2 (4), p. 5.

30. Bill Bryson (2004). *A short history of nearly everything*, p. 431, Broadway Books, New York.

31. I. Castles and D. Henderson (2003). Economics, Emission Scenarios and the Work of the IPCC, *Energy and the environment*, 14 (4), pp. 415–35.

32. *Ibid.*, pp. 425–7.

33. The Castles-Henderson critique is not confined to the SRES: it extends to the treatment of economic issues in the IPCC process generally. The IPCC's official response to their criticisms and proposals, in the form of a dismissive press release of December 2003, does not measure up to professional standards.

34. Robert Ehrlich (2003). *8 Preposterous Propositions*. Princeton and Oxford, p. 170. This book contains a useful summary of the main arguments on climate change, at pp. 138–87.

35. Even the link between a rise in global temperatures and a rise in sea levels is not beyond question. To offset the melting of glaciers and the thermal expansion of the oceans, a warmer world would probably mean more precipitation, more snow in Arctic regions, locking up more moisture, which would be removed from the oceans and would therefore lower sea levels. On the other hand, the current rise in sea levels is about twice what global warming levels would predict.

36. *Scientific American*, January, 2002.

37. One of these investigators, who had 'devoted his life to the subject', a leader of the anti-Lomborg inquisition, was Stephen Schneider. He had in fact devoted part of his earlier life to the advocacy of the view that we face a new ice age. In a paper written in the early 1970s, he argued that a vast increase in carbon dioxide emissions would have little warming effect, whereas the increase of aerosols in the atmosphere could well trigger an ice age. He continued to warn about a coming ice age until about 1978.

38. Philip Stott, *New Scientist*, 20 September, 2003.

39. S. Pimm and J. Harvey (2001). No need to worry about the future. *Nature*, **414**, 149.

40. Richard Stone in *Policy Summer*, Lomborg Review, 14 February, 2002.

41. Pimm and Harvey, *op. cit.*

42. M. Grubb, *Science*, **294**, 9 November, 2001, p. 1285.

43. A. Turner, 'Bjorn Again', *Prospect*, May, 2002, p. 28.

44. www.lomborg.org

45. *Spectator*, 23 February, 2002.

46. Rapid worldwide depletion of predatory fish communities. R. A. Myers and B. Worm, *Nature* 2003, Vol. 423, pp. 280–3.

47. 2003 IUCN Red List of Threatened Species. www.redlist.org. Downloaded 15 January, 2004.

48. Robert May, February 2002. '*Biological diversity in a crowded world: Past, present and future*', Blue Planet Prize, Asahi Glass Foundation, Tokyo.

49. Lomborg, *op. cit.*, pp. 255–6.

50. N. Myers and F. Lanting (1999). *International Wildlife*, **29** (2), 30–9.

51. C. D. Thomas, A. Cameron, R. E. Green, M. Bakkenes, L. J. Beaumont,

Y. C. Collingham, B. F. N. Erasmus, M. F. De Siqueira, A. Grainger, L. Hannah, L. Hughes, B. Huntley, A. S. Van Jaarsveld, G. F. Midgley, L. Miles, M. A. Ortega-Huerta, A. T. Peterson, O. L. Phillips and S. E. Williams (2004). Extinction risk from climate change. *Nature*, **427**, pp. 145–8. For a note of caution about the interpretation of these figures, see a letter from R. J. Ladle, P. Jepson, M. B. Araújo and R. J. Whittaker, *Dangers of crying wolf over risk of extinctions, Nature* **428**.

52. In his book *The Population Bomb*, published in 1968—admittedly a long time ago—Paul Ehrlich argued that India was a hopeless case, a country which could never feed its hungry, that there was no point in the rest of the world sending it aid.

53. www.lomborg.org

54. Turner, *op. cit.*, p. 335.

55. Spowers, *op. cit.*, p. 110. It is there claimed that this figure is rising at 7 per cent per year. A report in *The Guardian* of 5 March, 2004, gives a figure of 8 million.

56. See Prologue pp. 6–7.

57. Jay Byrne, '*Tactics and Tips*', September, 2003, AgBioView, www.agbioworld.org

CHAPTER SEVEN

1. *British Medical Journal* (2003). **327**, p. 694.

2. House of Lords, Science and Society, 2000, p. 22, para. 2.56.

3. Helene Guldberg, *Spiked Online*, 1 July, 2003.

4. *The Lancet* (2000). **356**, p. 265. This is a less qualified version of the Wingspread statement produced by a gathering of scientists, philosophers, lawyers and environmental activists in the United States in 1998, in which the final words 'not fully established scientifically' were used. See Guldberg, *op. cit.* In practice the qualification 'fully' tends to be left out.

5. European Environmental Agency, Earthscan (2002). 'The Precautionary Principle in the 20th Century, Late Lessons From Early Warnings'.

6. *Ibid.*, p. 31.

7. *Ibid.*, pp. 139–40.

8. John Emsley (1994). *The consumer's good chemical guide.* pp. 190–1, Oxford University Press.

9. EEA report, p. 50.

10. *Ibid.*, p. 58.

11. *Ibid.*, pp. 26–32.

12. Z. Jaworowski, Radiation Folly in *Environment and Health, Myths and Realities*, ed Okonski and Morris, International Policy Network, 2004.

13. An Introduction to Radiation Hormesis, S.M. Javad Mortazavi, www.angelfire.com/mo/radioadaptive/inthorm.html

14. *Ibid.*, p. 88.

15. *Ibid.*, p. 89.

16. The concept of 'unknown unknowns' has been strongly promoted by Professors Grove-White and Brian Wynne, both of Lancaster University and invoked more recently in a different context by Donald Rumsfeld.

17. Paul Brodeur, Currents of Death, quoted in Robert Park (2000), *Voodoo Science*, p. 153, Oxford.

18. *Ibid.*, p. 161.

19. For a comprehensive review of the scare about mobile phones and the lessons to be learned, see Adam Burgess (2003), *Cellular phones, public fears and a culture of precaution.* Cambridge.

20. *Ibid.*, p. 89.

21. EEA report, p. 210.

22. *Nature* (2002) **416**, p. 123.

23. Ray Tallis, *The enemies of hope*, p. 404, Macmillan, 1997.

24. Phillips, The BSE Inquiry, 2000, The Stationery Office.

25. *Ibid.*, Vol. 1, p. 266.

26. J. C. Hanekamp *et al*, Chloramphenicol, food safety and precautionary thinking in Europe, *Environmental Liability*, 11, 2003, 6 pp. 209–21.

27. WHO Pharmaceuticals Newsletter Nos. 7 & 8, July & August, 1999.

28. Lewis Smith (2001). *Trends in pharmacological sciences*, **22** (**6**), p. 281.

29. *New Scientist*, 17 May, 2003, p. 23.

30. Quoted from Ronald Bailey in Guldberg, *op. cit.*

31. Roger Allen (2001). Thalidomide regains respectability as new benefits are discovered. *The Pharmaceutical Journal*, **267**, p. 124.

32. Henry Miller and Gregory Conko, Policy Review no. 107.

33. John Kay (2003). *The truth about markets*, p. 174, Allen Lane.

34. www.Consumerfreedom.com, 2 April, 2003.

35. Frank Furedi, *Spiked Politics*, 15 March, 2002.

36. S. Wesseley, Psychological, social and media influences on the experience of somatic symptoms, September, 1997, quoted in B. Durodié

'Gender bending chemicals: facts and fiction'. www.spikedonline.com/
Articles/00000002D180.htm

CHAPTER EIGHT

1. Quoted in A. Sokal and J. Bricmont (2003). *Intellectual Impostures*, p. 250, Profile.
2. Quoted by Susan Haack in P. R. Gross, N. Levitt, and M. W. Lewis (eds.) (1996). *The flight from science and reason*. New York Academy of Sciences, p. 57.
3. House of Lords (2000). Select Committee on Science and Society and *See-through Science*, published by Demos, 2004.
4. Isaiah Berlin (1991). *The crooked timber of humanity*, p. 237, Fontana.
5. Bryan Magee (1973). *Popper*, p. 94, Fontana.
6. Karl Popper (1966). *The open society and its enemies*. Routledge and Kegan Paul.
7. Magee, *op. cit.*, pp. 96–7.
8. Quoted in A. Sokal and J. Bricmont, *op. cit.*, p. 181.
9. Ernest Gellner (1992). *Postmodernism, reason and religion*, p. 45, Routledge.
10. There is a passage in Heidegger's 1927 work *Sein und Zeit* ('Being and Time') on human existence, or being-there (*Dasein*), which even German readers might find it difficult to translate into ordinary German: '*Das Sein des Daseins besagt: Sich-vorweg-schon-sein-in-(der Welt) als Sein-bei (innerweltlich begegnendem Seienden).* Quoted by Mario Bunge in Gross, Levitt, and Lewis (1996). *The flight from science and reason*, p. 97.
11. Christopher Butler (2002). *Postmodernism*, p. 9, Oxford.
12. A. Sokal and J. Bricmont, *op. cit.*, Appendix A, pp. 199–240.
13. Tom Stoppard (1986). *Jumpers*, p. 30, Faber and Faber.
14. M. Douglas and A. Wildavsky (1983). *Risk and Culture*, pp. 80–1, California University Press.
15. House of Lords (2000). Select Committee on Science and Society, *op. cit.*, p. 24.
16. See-through Science, *op. cit.*, p. 18.
17. R. Fox, in Gross, Levitt, and Lewis (1996). *The flight from science and reason*, p. 330.
18. B. A. O. Williams (2002). *Truth and truthfulness*, p. 61, Princeton University Press.
19. House of Lords, *op. cit.*, p. 24.

20. Stephen Cole, in Gross, Levitt, and Lewis (1996). *The flight from science and reason*, p. 275.

21. J. Radcliffe Richards, *Ibid.*, p. 385.

22. G. G. M. James (1954). *Stolen Legacy*. New York.

23. M. Lefkowitz, in *The flight from science and reason*, *op. cit.*, pp. 304–6.

24. M. Bunge, *Ibid.*, p. 106.

25. G. Holton, *Ibid.*, p. 558.

26. J. Meek, *The Guardian*, 4 February, 2002.

27. House of Lords, *op. cit.*, pp. 26–7.

28. The Royal Society 'Genetically modified plants for food use and human health—an update', February, 2002.

29. Tony Gilland, www.spiked-online.com, 14 February, 2002.

30. RS Policy document 19/04, July, 2004.

31. Annual Report, Food Standards Agency, 2003.

32. See Bill Durodié, *Limitations of public dialogue in science and the rise of new experts*, *Critical Review of International Social and Political Philosophy*, 15, No. 4, p. 86ff.

33. Tallis, *op cit.*, p. 199.

34. House of Lords, *op. cit.*, p. 22, para 2.56.

35. C. Hood, H. Rothstein, and R. Baldwin (2001). *The government of risk*, p. 102, Oxford.

36. Demos, *See-through science*, *op. cit.*, see pp. 26, 28, 29, 62.

37. A valuable essay on this topic is Bill Durodié's, 'The Demoralisation of Science', p. 13, paper presented to conference on 'Demoralization: Morality, Authority and Power' held at cardiff University, 5–6 April 2002, available at: http://www.cf.ac.uk/sosci/news/dmap/papers/Durodie.pdf

38. N. Koertge, in Gross, Levitt, and Lewis (1996). *The flight from science and reason*, p. 270.

39. Michael Shermer (2001). *The borderlands of science*. pp. 87–8, Oxford.

40. Onora O'Neill (2002). *Autonomy and truth in bioethics*, p. 169, Cambridge.

41. Marcia Angell (1997). *Science on trial*. Norton. One plaintiff recovered US$7.34 million in damages (p. 55).

CHAPTER NINE

1. *Radical Economics*, 13, July/August, 2003.

2. One of these conspiracy theorists, George Monbiot, writes a regular

column in *The Guardian,* plainly connected to the Gmwatch.org website that posts full details of all the interconnecting webs of pro-GM forces linked by corporate funds of some kinds.

3. Will Hutton (2003). *The world we're in,* p. 31, Abacus.

4. John Plender (2003). *Global capital and the crisis of legitimacy.* John Wiley.

5. Jagdish Bhagwati, *In defence of globalisation,* p. 168, Oxford.

6. Jon Thompson, Patricia Baird, and Jocelyn Downie (2001). *The Oliveiri Report.* James Lorimer.

7. Philippe Legrain 'Against Globaphobia', *Prospect,* May, 2000, p. 31. See also Legrain (2002). *Open world: The truth about globalisation,* p. 50, Abacus.

8. Martin Wolf, *Why globalisation works,* p. 141, Yale.

9. *Ibid,* p. 308.

10. Kofi Annan and Henry Louis Gates Jr. 'On Africa, the UN and Globalization', *Prospect,* July, 2001, p. 19.

11. http://people-press.org/reports/display.php3?ReportID=165

12. House of Lords Select Committee on Economic Affairs, Globalisation, 2002, p. 19. For a full discussion of different figures for the decline in world poverty, see Martin Wolf, *op. cit.,* pp. 157–63.

13. Naomi Klein (2001). *No Logo,* p. 37, Flamingo.

14. Bhagwati, *op. cit.,* p. 152.

15. Commission on Intellectual Property Rights, DFID, 2003, p. 21.

16. *Ibid.,* p. 36.

17. *Ibid.,* p. 32.

18. Shereen El Feki 'Drugs for the World's Poor', *Prospect,* November 2000, p. 42.

19. DFID, *op. cit.,* p. 43.

20. Klein, *op. cit.,* pp. 339–40.

21. Noreena Hertz (2002). *The silent takeover.* Heinemann.

22. Wolf, *op. cit.,* pp. 221–3; Bhagwati, *op. cit.,* p. 166.

23. Dick Taverne (1999). *Tax and the Euro,* Centre for Reform, p. 22. For full details of tax rises in OECD countries, see Wolf, *op. cit.,* pp. 253–7.

24. Klein, *op. cit.,* p. 486.

25. Philippe Legrain (2002). *Open world: The truth about globalization.* pp. 56–62, Abacus.

26. Wolf, *op. cit.,* p. 236.

27. Legrain, *op. cit.,* p. 62.

28. ILO Report (2002), 'A Future Without Child Labour', p. 16.

29. House of Lords Committee, *op. cit.,* Evidence II Q1246.

30. *Ibid.*, p. 27.

31. Richard Tomkins, *Financial Times*, 6 June, 2003.

32. For this description, I am indebted to Dennis Dutton, who teaches philosophy at the University of Canterbury, New Zealand.

33. John Vidal, *The Guardian*, 11 November, 2002.

34. Naomi Klein, *The Guardian*, 1 February, 2003.

35. *Ibid.*

36. Martin Wolf 'Klein's Clangers', *Prospect*, February 2003, p. 74.

37. The cry 'Take to the streets' was proclaimed by the columnist George Monbiot in *The Guardian* (2 March, 2004) and anti-GM activists advocate 'Trash the crops' as a response to the official licensing of GM maize in the UK.

38. Naomi Klein, *The Guardian*, 9 November, 2002.

39. *Ibid.*

40. John Kay (2003). *The truth about markets*, p. 323, Allen Lane.

41. *Ibid.*, p. 8.

42. Hutton, *op. cit.*, p. 26.

CHAPTER TEN

1. Quoted in Lesley Stephen (1962). *History of English thought in the Eighteenth Century, vol. 1*, p. 197, Harbinger.

2. A. Sokal and J. Bricmont (2003). *Intellectual impostures*, p. 191, Profile.

3. In the Presidential election of 2000 in the USA, 45 per cent of the electorate voted. In Britain in the General Election of 2001, 58 per cent voted, much the lowest turn-out in a general election since World War II. However, on the continent of Europe, where the electoral system is one of proportional representation that makes each individual vote count for more, turn-out in elections is much higher.

4. Thomas Patterson, Out of Order, quoted in John Lloyd, *Prospect*, October, 2002, p. 49.

5. *Ibid.*, p. 52. Since this book was written, John Lloyd has published *What the Media are doing to our Politics*, Constable and Robinson, 2004.

6. John Stuart Mill (1859). *On liberty and other essays* (Oxford University Press edn., 1991).

7. I regard *Zen* as one of the great, bad books of its time. It operates at a number of levels, as a moderately readable novel about a man who may commit suicide and his relations with his son, as a travelogue about part

of America, and as a rather unconvincing discussion of the philosophy of Sophism and Zen. The most interesting part for me was about motorcycle maintenance. It altered my attitude towards engines. Instead of reacting with fury when the outboard on my dinghy did not work, which made me want to throw it overboard, I realized there was an intellectual problem capable of intelligent solution even by those with limited technical knowledge like myself.

8. Karl Popper (1966). *The open society and its enemies*. Routledge and Kegan Paul.

9. The argument in the paragraph that follows is a summary of the relevant part of Bryan Magee's excellent booklet, *Popper* (1973), pp. 74–7.

10. The twenty most prosperous nations in the world are all democracies. One apparent exception to the rule that freedom plays an important part in enabling economic growth is China, where economic liberalisation and rapid growth has not been accompanied by more democracy. However, the Chinese economy has been extensively decentralised. It seems likely that this will make autocratic political control increasingly difficult. The government is also concerned about the widespread corruption that is a threat to sustained growth. It is unlikely that corruption can be brought under control without the relaxation of political dictatorship.

11. P. Dasgupta (2001). *Human well-being and the natural environment*, p. 75, Oxford.

12. John Kay (2003). *The truth about markets*, p. 88, Allen Lane.

13. Walter Gratzer (2000). *The undergrowth of science*. pp. 177ff., Oxford.

14. *Ibid.*, pp. 177–204.

15. *Ibid.*, pp. 204–12.

16. *Ibid.*, pp. 219ff.

17. Quoted in G. Holton (1993). *Can science be at the centre of modern culture? Public understanding of science, vol. 2*, p. 302.

18. Mario Bunge, in Gross, Levitt, and Lewis (1996). *The flight from science and reason*, p. 110, New York Academy of Sciences.

19. Quoted in B. A. O. Williams (2002). *Truth and truthfulness*, p. 144, Princeton University Press.

20. Gratzer, *op. cit.*, p. 244.

21. *Ibid.*, p. 301.

22. *Ibid.*, p. 226.

23. Jason Burke (2003). *Al-Qaeda*, p. 14, Tauris.

24. Douglas Adams, quoted in Richard Dawkins (2003). *A devil's chaplain*, p. 156, Weidenfeld and Nicolson.

25. *The Guardian*, 25 November, 2002.

26. Victoria Clark 'The Christian Zionists', *Prospect*, July, 2003, p. 54.

27. *Ibid.*, p. 55.

28. *The Economist*, 8 February, 2003, p. 55.

29. Don Bruce and Don Horrocks (eds.) (2001). *Modifying Creation*. Paternoster Press.

30. House of Lords, Official Report, 2002, Vol. 621, col. 35–6.

31. Sokal and Bricmont, *op. cit.*, p. 191.

32. Bernard Lewis (2002). *What went wrong? The clash between Islam and modernity in the Middle East.* p. 174, Penguin.

33. *Ibid.*, p. 8.

34. *The Economist*, Special Survey of Islam, 13 September, 2003, p. 6.

35. Vicenzo Oliveti (2003). *Terror's Source*, p. 10ff., Amadeus.

36. *Ibid.*, p. 5.

37. *Ibid.*, p. 7.

38. Bernard Lewis, *op. cit.*, p. 51.

39. *Ibid.*, p. 113.

40. Burke, *op. cit.*, p. 38.

41. Oliveti, *op. cit.*, p. 8.

42. Burke, *op. cit.*, p. 38.

43. MORI poll, March 2001.

Index